AI Knowledge Transfer from the University to Society

AI Knowledge Transfer from the University to Society

Applications in High-Impact Sectors

Edited by
José Guadix Martín
Milica Lilic
Marina Rosales Martínez

CRC Press
Taylor & Francis Group
Boca Raton London New York

CRC Press is an imprint of the
Taylor & Francis Group, an **informa** business

First Edition published 2022
by CRC Press
6000 Broken Sound Parkway NW, Suite 300, Boca Raton, FL 33487-2742

and by CRC Press
4 Park Square, Milton Park, Abingdon, Oxon, OX14 4RN

CRC Press is an imprint of Taylor & Francis Group, LLC

© 2022 selection and editorial matter, José Guadix Martín, Milica Lilic, Marina Rosales Martínez; individual chapters, the contributors

Reasonable efforts have been made to publish reliable data and information, but the author and publisher cannot assume responsibility for the validity of all materials or the consequences of their use. The authors and publishers have attempted to trace the copyright holders of all material reproduced in this publication and apologize to copyright holders if permission to publish in this form has not been obtained. If any copyright material has not been acknowledged please write and let us know so we may rectify in any future reprint.

Trademark notice: Product or corporate names may be trademarks or registered trademarks and are used only for identification and explanation without intent to infringe.

Library of Congress Cataloging in Publication Data
A catalog record has been requested for this book.

ISBN: 978-1-032-22632-3 (hbk)
ISBN: 978-1-032-23301-7 (pbk)
ISBN: 978-1-003-27660-9 (ebk)

DOI: 10.1201/9781003276609

Typeset in Times
by Deanta Global Publishing Services, Chennai, India

This book has been published within the framework of the unique project "Innovative Ecosystem with Artificial Intelligence for Andalusia 2025", subsidized by the Autonomous Government of Andalusia and coordinated by the AndalucíaTech International Campus of Excellence, where its two universities, Seville and Malaga, work with technology companies for the development of Artificial Intelligence technologies in different sectors. The book focuses solely on the subprojects carried out at the University of Seville.

Contents

Foreword

Artificial Intelligence: The New Paradigm to Boost Society 5.0

Artificial Intelligence (AI) takes human intelligence and bio-inspired systems as a reference and has application to numerous fields of society and science. The applicability and great transformation potential of AI rely on its deployment in manufacturing environments, with Industry 4.0 as an exponent, its application in the fields of biomedicine and reliable economic transactions, through blockchain approaches, or the identification of social profiles, among others very diverse examples. All this makes Artificial Intelligence a great success story told in the first decades of the current 21st century.

Indeed, although Artificial Intelligence, as a field of work, dates from the second half of the last century, its impulse has occurred in the first decades of the 21st century. The reason is mainly based on:

- The huge availability of unstructured data that needs to be analyzed (Big Data)
- Today's high-capacity and low-cost computing systems with computing capabilities that are already measured in petaflops
- The development and progress of complex Artificial Intelligence structures, associated, for instance, with deep learning, hybridization of fuzzy networks, or the development of Machine Learning

There is no doubt that one of the main challenges that today's society faces is to ensure the incorporation of AI technology throughout its economy.

ARTIFICIAL INTELLIGENCE IN THE GLOBAL AND EUROPEAN CONTEXT

Most of the world's major economies are indisputably betting on the promotion of Artificial Intelligence and its application to all fields of society. Thus, the US Government has already in 2016 invested close to 970 million euros in research on AI technologies. Since then, the country has been increasing its economic contribution to the development of the field. On the other hand, with its "Next Generation Artificial Intelligence Development Plan", China has committed to achieving world leadership in 2030 and is investing heavily. Also, other countries, such as Japan and Canada, have adopted strategies in relation to AI.

In the European context, for the period 2018–2020, the Commission determined that it would invest around 1.5 billion euros in promoting research and innovation in AI technologies, enhancing AI applications that can provide solutions to social challenges, supporting cutting-edge initiatives in the field, and creating centers of excellence specialized in AI research. Also, the Commission established the need to attract more private investment in AI through the European Fund for Strategic Investments.

The European Commission proposal for the scenario that arises from 2020, within the current multiannual financial framework 2021–2027, consists of several goals, including:

- The improvement of the pan-European network of centers of excellence specialized in AI
- Research and innovation in areas such as explainable AI, unsupervised Machine Learning, and energy and data efficiency
- Additional digital innovation poles, state-of-the-art testing, and experimentation facilities in areas such as health care, transportation, the agrifood industry, and the manufacturing sector
- The creation of a data-sharing support center, which will be closely linked to the AI-on-demand platform and whose mission will be to facilitate the development of applications for companies and the public sector

ARTIFICIAL INTELLIGENCE IN THE SPANISH AND THE ANDALUSIAN CONTEXT

Within the Framework of the Preparation of the Spanish Strategy for Science, Technology and Innovation, the Ministry of Science and Innovation has drawn up different sectoral strategies, among which the Spanish Strategy for R&D in Artificial Intelligence acquires an indisputable role, which allowed the development of the National Artificial Intelligence Strategy (ENIA in Spanish). The ENIA has six strategic axes:

- Promotion of scientific research, technological development, and innovation in Artificial Intelligence
- Promotion of digital skills, the development of national talent, and the attraction of international talent
- Development of data platforms and technological infrastructures that support AI
- Integration of AI into value chains to transform the economic market
- Promotion of the use of AI in the Public Administration and in national strategic missions
- Establishment of an ethical and regulatory framework that guarantees the protection of individual and collective rights, with social welfare and sustainability

When it comes to Andalusia, the region located in the south of Spain, the current framework that defines the Andalusian R&D Strategy, together with the Smart Specialization Strategy for the Sustainability of Andalusia for the period 2021–2027, called S4, will set the guidelines for boosting the Artificial Intelligence application in the region.

Andalusia already has a remarkable position in the field of work. At the universities of Seville, Granada, Malaga, and Jaén, there is a very relevant group of researchers who carry out their scientific activities in the field of AI and who occupy prominent positions in the world rankings of scientific impact (highlighting the Academic Ranking of World Universities [ARWU] ranking or that of Stanford University, among others).

As a final consideration, the following three aspects should be underlined: (i) the enormous transformative potential of Artificial Intelligence for society in all its fields of development, (ii) the economic transformation that has taken place in the business world, since seven of the ten companies with the largest world market capitalization in 2020 were already users or developers of this technology, and (iii) the mandatory responsibility of public administrations to collaborate to promote the Artificial Intelligence at the different levels of society, on the one hand, and to establish the necessary ethical framework for its development, on the other.

Pablo Cortés Achedad
General Secretary for Business, Innovation and Entrepreneurship
Autonomous Government of Andalusia, Spain

Editors

José Guadix Martín, PhD in Industrial Engineering, Full Professor, Department of Industrial Organization and Business Management II, University of Seville, Seville, Spain.

Milica Lilic, PhD in Languages, Texts and Contexts, Entrepreneurship and R&D Project Manager at the Secretariat of Knowledge Transfer and Entrepreneurship, University of Seville, Seville, Spain.

Marina Rosales Martínez, PhD in Industrial Engineering, Head of Service at the Secretariat of Knowledge Transfer and Entrepreneurship, University of Seville, Seville, Spain.

Contributors

Orly Enrique Apolo-Apolo
Aerospace Engineering and Fluid
    Mechanics Department
Agroforestry Engineering Area
University of Seville
Seville, Spain

Pedro Blanco Carmona
Researcher under contract
Department of Electronics
    Engineering
Technical School of Engineering
University of Seville
Seville, Spain

Manuel Castro Malet
4i Intelligent Insights S.L.
Seville, Spain

Pablo Cortés Achedad
General Secretary for Business,
    Innovation and Entrepreneurship
Autonomous Government of
    Andalusia
Spain

Adolfo Crespo Márquez
Full Professor
Department of Industrial
    Management
Technical School of Engineering
University of Seville
Seville, Spain

Antonio De la Fuente Carmona
Pre-Doctoral Researcher
Department of Industrial
    Management
Technical School of Engineering
University of Seville
Seville, Spain

Ricardo Durán Viñuelas
Julietta Research Group in Natural
    Language Processing
University of Seville
Seville, Spain

María-José Escalona
Full Professor
Department of Computer Systems
    and Languages
Technical School of Computer
    Engineering
University of Seville
Seville, Spain

Antonio Estepa Alonso
Associate Professor
Department of Telematics
    Engineering
Technical School of Engineering
University of Seville
Seville, Spain

Rafael Mª Estepa Alonso
Associate Professor
Department of Telematics
  Engineering
Technical School of Engineering
University of Seville
Seville, Spain

Rafael Fernández-Chacón
Full Professor
Department of Medical Physiology
  and Biophysics and CIBERNED
Institute of Biomedicine of Seville
  (IBiS, University Hospital Virgen
  del Rocío/CSIC/University of
  Seville)
Seville, Spain

Pablo Fernández Montes
Associate Professor
Department of Computer Languages
  and Systems
Technical School of Computer
  Engineering
University of Seville
Seville, Spain

José María García
Associate Professor
Department of Computer Languages
  and Systems
Technical School of Computer
  Engineering
University of Seville
Seville, Spain

Francisco A. Garcia Benitez
Full Professor
Department of Transport
  Engineering and Infrastructure
Technical School of Engineering
University of Seville
Seville, Spain

Antonio Garcia-Martinez
Assistant Professor
University Institute of Architecture
  and Construction Sciences
Technical School of Architecture
University of Seville
Seville, Spain

José Ramón García Oya
Associate Professor
Department of Electronics
  Engineering
Technical School of Engineering
University of Seville
Seville, Spain

Emilio Gomez-Gonzalez
Full Professor of Applied Physics
Department of Applied Physics III
School of Engineering
University of Seville
Seville, Spain

Ramón González Carvajal
Full Professor
Department of Electronics
  Engineering
Technical School of Engineering
University of Seville
Seville, Spain

Nuria Gómez-Vargas
Institute of Mathematics of the
    University of Seville (IMUS)
Seville, Spain

Yolanda Hinojosa Bergillos
Associate Professor
Department of Applied Economic I
    and Institute of Mathematics
University of Seville
Seville, Spain

**Francisco José Jiménez-Espadafor
Aguilar**
Full Professor
Department of Energy Engineering
Technical School of Engineering
University of Seville
Seville, Spain

Diego Francisco Larios Marín
Assistant Professor
Politecnic School
University of Seville
Seville, Spain

Fernando Lazcano Alvarado
Department of Transport
    Engineering and Infrastructure
Technical School of Engineering
University of Seville
Seville, Spain

Juan A. Leñero-Bardallo
Associate Professor
Institute of Microelectronics of
    Seville (IMSE-CNM)
University of Seville and State
    Agency for the Higher Council
    for Scientific Research (CSIC)
Seville, Spain

Javier Marquez-Rivas
Staff Neurosurgeon
Service of Neurosurgery
University Hospital 'V. Rocio'
Seville, Spain

Pedro Martín-Holgado
National Accelerators Center (CNA)
University of Seville/Spanish
    National Research Council
    (CSIC)/Autonomous Government
    of Andalusia
Seville, Spain

Francisco Javier Molina Cantero
Assistant Professor
Politecnic School
University of Seville
Seville, Spain

Yolanda Morilla García
National Accelerators Center (CNA)
University of Seville/Spanish
    National Research Council
    (CSIC)/Autonomous Government
    of Andalusia
Seville, Spain

Jesús Muñuzuri Sanz
Full Professor
Department of Industrial
    Organization and Business
    Management II
Technical School of Engineering
University of Seville
Seville, Spain

José Navarro-Pando
President of the Inebir Group
Director of the Human Reproduction
    Unit in Seville
Seville, Spain

Luis Onieva Giménez
Full Professor
Organization Engineering research
   group
University of Seville
Seville, Spain

Juan Pedro Pérez Alcántara
Department of Physical Geography
   and Regional Geographic
   Analysis
Faculty of Geography and History
University of Seville
Seville, Spain

Manuel Pérez-Ruiz
Associate Professor
Aerospace Engineering and Fluid
   Mechanics Department
Agroforestry Engineering Area
University of Seville
Seville, Spain

Francisco Pinto Puerto
Full Professor
Department of Architectural Graphic
   Expression
Technical School of Architecture
University of Seville
Seville, Spain

Diego Ponce López
Postdoctoral Fellow
Department of Statistics and
   Operations Research and Institute
   of Mathematics
University of Seville
Seville, Spain

Gonzalo Quirosa Jiménez
Energy Engineering Department
Technical School of Engineering
University of Seville
Seville, Spain

Jasone Ramírez-Ayerbe
Institute of Mathematics of the
   University of Seville (IMUS)
Seville, Spain

Alicia Robles-Velasco
Pre-Doctoral Researcher
Department of Industrial
   Organization and Business
   Management II
Technical School of Engineering
University of Seville
Seville, Spain

Ángel Rodríguez-Vázquez
Full Professor
Institute of Microelectronics of
   Seville (IMSE-CNM)
University of Seville and State
   Agency for the Higher Council
   for Scientific Research (CSIC)
Seville, Spain

Esperanza Sánchez Rodríguez
Department of Physical Geography
   and Regional Geographic
   Analysis
Faculty of Geography and History
University of Seville
Seville, Spain

Miguel Torres García
Full Professor
Department of Energy Engineering
Technical School of Engineering
University of Seville
Seville, Spain

Antonio Jesús Torralba Silgado
Full Professor
Department of Electronics
  Engineering
Technical School of Engineering
University of Seville
Seville, Spain

Marina Valenzuela-Villatoro
Institute of Biomedicine of Seville
  (IBiS, University Hospital Virgen
  del Rocío/CSIC/University of
  Seville)
Department of Medical Physiology
  and Biophysics and CIBERNED
Spain

Juan Manuel Vozmediano Torres
Associate Professor
Telematics Engineering Department
Technical School of Engineering
University of Seville
Seville, Spain

Introduction

In the early days of the Spanish public university, only teaching activities were carried out, since this institution was considered a place where students acquired knowledge which enabled them to develop a professional activity. Subsequently, teachers with inquisitiveness began to make some progress in the discovery of new knowledge, going beyond what was explained in the classroom. These were the beginnings of research work, which in recent years has become widespread among most professors.

Moreover, in the past 40 years, some of these professors and researchers began to transfer the knowledge acquired in their research to institutions and companies in the surrounding area that required it. In this way, the productive sector has gained a crucial element for its innovation in products and services, not only formed by the R&D departments of large corporations but also supported by university research groups, with the knowledge of their staff and laboratory facilities. This contracted innovation is known as knowledge transfer, and in Spain, it is included as the third mission of the university in the Organic Law on Universities and in the Law on Science, Technology, and Innovation of 2011.

As a result of this relationship between teaching and research staff and institutions and companies, university staff begin to complement their research work with the needs of society. It raises its concerns and researchers begin to think about how to solve them. This is mostly reflected in institutions and companies that have to improve specific aspects or acquire new processes that they currently lack. In this way, the possibilities of our society are improved, by creating jobs with high added value, requiring qualified staff, and having a competitive advantage in the companies that collaborate with the university. During this bidirectional process, innovation and knowledge transfer take place from the heart of the universities to the environment.

Here we find a crucial characteristic of innovation, which is the concern to solve a problem in society or to improve the current situation. Therefore, there are researchers who also seek to anticipate customer needs and solve future problems.

Traditionally, it has been thought that university innovation and knowledge transfer were research contracted with institutions, through the contracts of the articles of the Spanish Organic Law on Universities 68/83, intellectual property, etc. However, these are just tools. In university innovation, it is an essential requirement to have agile processes and tools that allow the knowledge generated in the university to be transmitted, without the professors dying in a tangle of administrative tasks. Likewise, it is not a reactive activity but a proactive one. As it is a function of the Spanish public university, waiting for knowledge to be requested is not an option. On the contrary, it is assumed that work must be done to find those who can use it or to facilitate the creation of innovative companies.

Having detected this set of activities associated with university innovation and knowledge transfer, the University of Seville was a pioneer in Spain, creating a specific Vice-Rector's Office for this purpose in 2004, the Vice-Rector's Office for Technology Transfer. Recently, in the first call for proving the results of six-year knowledge transfer in 2018, published by the Ministry of Education and Vocational Training, the University of Seville has been the Spanish university with the most professors who have achieved this recognition.

Among the most important actions during almost 20 years of promoting knowledge transfer, the Research Foundation of the University of Seville stands out, as an entity that provides an administrative structure that adjusts university innovation to the pace of business, as well as the company chairs, that is, company donation agreements that help deepen in new research areas of interest to society. In addition, innovative entrepreneurship is encouraged, boosting the capacities and creation of companies. Also, it is worth mentioning that the University of Seville is a global partner of the University of California-Berkeley since 2016.

Following the same line of innovation and knowledge transfer, this book has arisen as a result of a novel action that brings together examples of the progress achieved by the University of Seville in the unique project "Innovative Ecosystem with Artificial Intelligence for Andalusia 2025". It has joined a total of 22 subprojects, made of research groups, on the one hand, and different institutions and/or companies that act as aggregate agents, on the other, with the common denominator of using Artificial Intelligence to incorporate innovation in their lines of action and help them in decision-making. The relevance of Artificial Intelligence, its interdisciplinary nature, and its ability to accelerate global solutions have turned this set of technologies into a tool aimed at generating a new social and economic reality. Hence, the transformative power of AI in the areas of health, sustainability, infrastructure, security, and tourism, among others, has motivated the integration of these disruptive technologies in all the subprojects that are the subject of this book.

Thus, starting from different Artificial Intelligence techniques, such as Machine Learning, Blockchain, Internet of Things, Big Data, Business Analytics, Robotics, Cybersecurity, Augmented Reality, Building Information Modeling (BIM), and Geographic Information System (GIS), the researchers involved in the mentioned unique project, together with collaborating

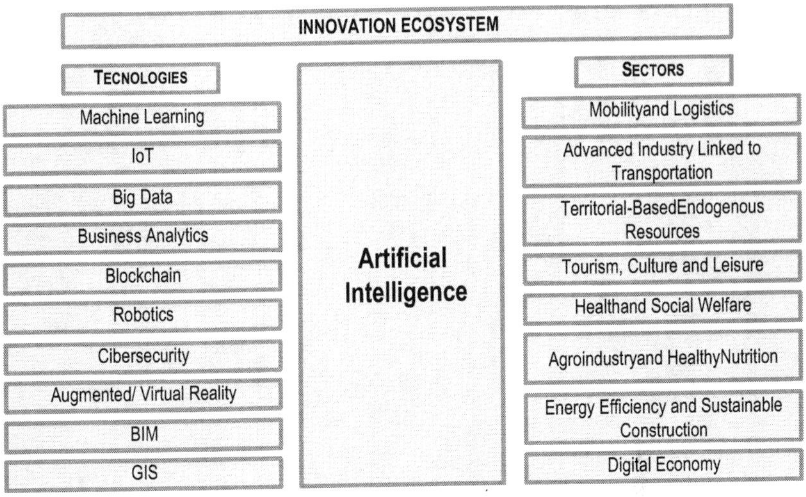

FIGURE I.1 Artificial Intelligence techniques and sectors involved in the project.

institutions and/or companies, have studied different ways of taking advantage of technological knowledge in the following sectors: health and social welfare; renewable energies, energy efficiency, and sustainable construction; advanced industry linked to transport; mobility and logistics; endogenous resources with a territorial base; tourism, culture, and leisure; agribusiness and healthy nutrition; digital economy (see Figure I.1).

These eight sectors, due to thematic similarity, will be collected in five chapters of this book, where some of the progress of these 22 bidirectional lines of action (university–company) will be specified. The current situation of the field in which specific research operates, on the one hand, and its novel contribution, on the other hand, will be presented.

Therefore, apart from Artificial Intelligence as the common thread of this book, its relevance is reflected in the set of aggregate agents selected for the project, that is, institutions, corporations, and medium-sized companies, with an outstanding dynamism component, which will allow an appropriate technology transfer to society.

This research is aligned with the Spanish Strategy for Science, Technology and Innovation 2021–2027, within the Government Delegate Commission on Science, Innovation and Universities (28/12/2018), which drew up the R&D Sectoral Strategies, and this proposal is framed in the Spanish Strategy for R&D in Artificial Intelligence.

The coordinators of this work hope that the use of Artificial Intelligence technologies can be a leverage that helps the digital transformation and achieving more efficient and productive companies that result in improvements for the region of Andalusia in Spain.

José Guadix Martín
Milica Lilic
Marina Rosales Martínez
(Eds.)

Health and Social Welfare

1

The health and well-being of citizens is a key factor that must be guaranteed, as has recently become clear with the global health pandemic. Being aware of this, the University of Seville research groups apply Artificial Intelligence techniques to address existing problems from a multidisciplinary perspective and improve current solutions. The health sector has large amounts of patient data, which, when correctly analyzed and visualized, will increase the efficiency of the applied treatments. For this reason, there are medicine, computer science, or physics groups that address different issues.

An example of this is the application of Artificial Intelligence to optimize focused ultrasound cleaning of implanted shunts in patients with different pathologies. An innovative technology is used for the non-invasive preventive cleaning of shunts, valves, or catheters implanted in patients with different pathologies. Artificial Intelligence on the data of test cases determines the optimal parameters for its application in diverse clinical environments, adapted to the individual circumstances and the specific devices used in neurosurgery, oncology, and other clinical areas.

Likewise, another example of robotics that allows automating medical processes is shown, guaranteeing the quality and safety of the steps carried out during the process. It is intended to improve health processes with the use of emerging technologies, such as blockchain or the robotization of software processes.

Furthermore, there is another situation generated by the extension of life expectancy in modern societies and the social challenge that arises in the prevention and treatment of neurodegenerative diseases suffered by the constantly increasing elderly population. The underlying problem is the investigation of changes in gene expression of individual neurons in response to nerve terminal dysfunction in the brain of genetically modified mice. Within this experimentation, Machine Learning methods are used for the bioinformatic analysis of a data set that includes thousands of cells with thousands of genes each. This issue has a utility to be addressed in future therapeutic strategies.

DOI: 10.1201/9781003276609-1

1.1 ARTIFICIAL INTELLIGENCE FOR THE OPTIMIZATION OF FOCUSED ULTRASOUND CLEANING OF SHUNTS IMPLANTED IN PATIENTS OF DIFFERENT PATHOLOGIES

Emilio Gómez-González and Javier Márquez-Rivas

ABSTRACT

Focused ultrasound is an innovative technology for potential non-invasive, preventive cleaning of shunts and infusion systems (valves, catheters) implanted in patients with different pathologies. Artificial Intelligence tools are used to determine optimal parameters for their application in various clinical settings, tailored to the individual circumstances and the specific devices used in neurosurgery, oncology, and other clinical areas.

INTRODUCTION

The expanding implementation of technologies based on Artificial Intelligence (AI) represents an authentic revolution in most aspects of daily life, having a very relevant impact on Medicine and Healthcare (Gómez González & Gómez, 2020). Diagnosis and treatment methods powered by AI, and their combinations with augmented reality (AR) devices, evolve toward personalized three-dimensional (3D) models, and to the development of advanced simulation tools for predictive, interactive visualization of the response of organisms under different circumstances and stimuli.

The European Union considers the development of AI an essential area, particularly in Medicine and Healthcare, and has launched different initiatives to lead its scientific and technological advances and to establish the bases for their implementation (Nepelski, 2021). One of these initiatives is the ATTRACT Program, which seeks to identify and support groundbreaking technologies with a clear potential of application for solving societal, clinical, environmental, problems of high impact (ATTRACT Program, 2019). This work presents an extension of a successful project within the aforementioned ATTRACT Program (FUSCLEAN Project, 2020).

STATE OF THE ART

The project described in this article addresses the preventive cleaning of fluid infusion or drainage (shunting) systems implanted in the human body to avoid their obstruction by the deposit of residues on their inner surfaces.

Intended cleaning is achieved by means of focused ultrasound beams that generate a certain 3D distribution of energy, safe for the patient but which produces mechanical effects—controlled cavitation—in the fluid which, in turn, disaggregate and remove the deposits from the walls of the conduits and valves.

The initial field of application of this technology is neurosurgery, to avoid the most frequent complications in patients with hydrocephalus, but it has the potential to be extended for additional applications in other clinical areas in which shunting systems are implanted in the human body to infuse or drain fluids, such as pain control, oncology treatments, and anesthesia.

In this research, AI systems are employed to conduct multiple 3D simulations to determine and optimize the parameters required to achieve the desired level and spatial distribution of ultrasound energy volume density in the targeted regions. AR systems are also tested. They include different types of display, from screens or tablets to glasses or helmet-mounted devices, to superimpose "layers" of information on the user's real field of vision (Baraas et al., 2021). They can also be integrated with positioning and motion tracking elements to enable simultaneous viewing of real objects with data or graphics in different orientations or perspectives.

Currently, there are no references to early detection of debris in fluids flowing through implanted shunts in the human body or to preventive procedures to avoid the deposits of materials in the conduits and valves and their possible obstruction. Therefore, no additional references are found about energy focusing and visualization technologies for the intended cleaning or about similar applications.

CONTRIBUTION

The described technology is based on the generation and real-time control of a volumetric distribution of ultrasound energy in the region of interest (called "sonication volume"), during the time interval necessary to produce the detachment of deposits adhered to the inner surfaces of conduits and valves. It is important to note that ultrasound waves are mechanical waves (not electromagnetic radiation), without ionizing effects. The intended effect is achieved by concentrating individual beams in the sonication volume, analogously to the concentration of the sun's rays using a magnifying glass.

The desired energy distribution is achieved by the overlapping of the ultrasound fields generated by independent emitters at certain frequencies, whose powers and orientation are precisely determined. This process requires identifying the exact location of the shunt elements—catheters and valves—under the skin and generating a 3D numerical model of the emission of the various ultrasound beams. The combined ultrasound field produces a controlled cavitation effect in the targeted volume—inside the conduits—that disintegrates the debris in the deposits so that they are evacuated by the circulation of the fluid itself. Likewise, it is necessary to monitor the temperature distribution in the zone surrounding the region of concentration of the ultrasound energy to avoid possible thermal effects. We, therefore, explore the combined use of visible and thermal imaging cameras.

In this technology, AI algorithms are used to optimize the emission patterns of the individual ultrasound beams, considering the various factors inherent to the implanted systems (types of valves and conduits, with their different mechanical properties), the circulating fluid (biofluids, drugs) and the patient. It is also required to include in the calculations the biological variability of elements which may be present in fluids and the state of the corresponding system of each person, as well as their specific conditions (temperature, cardiac and respiratory pulsations, possible concurrent pathologies) at the time of the application of the ultrasound beams. The different types, positions, and circumstances of the implanted valves and conduits must also be considered since the presence of elements such as scar tissue, and the various characteristics of the layers of the skin and subcutaneous fat, determine the effective distribution of ultrasound in the targeted structures, that is, within the implanted devices and components. This multitude of factors is completely different for each clinical setting and person. Multiple 3D simulations must be developed in a variety of physical scenarios, under varying circumstances and with different materials of complex properties. From them, the optimal sonication parameters are extracted using AI tools.

In our work, we explore clinical applications in which the consequences of possible obstructions of implanted systems may be particularly relevant for patient safety and the effectiveness of the treatment, and in which the removal and replacement of the implanted devices are more complex or difficult. These areas of applications include, in addition to patients of hydrocephalus in neurosurgery, treatments based on drug infusion for pain control and chemotherapy in oncology and anesthesia. Such developments are possible within the institutional collaboration framework of this project, with the participation of aggregate agents that complement from the practical, medical, and clinical points of view, the scientific analysis from the academic field.

From a technical perspective, this project is based on the availability of AI tools that allow the optimization of the simulations of the different scenarios and visualization systems using augmented reality devices. AI algorithms mainly belong to the "Machine Learning" type. In general, although they need extended data sets for training, they have a great advantage—for subsequent practical implementation—of their ability to improve by increasing the number and typology of processed cases. We also explore different AR devices to allow for intuitive, user-friendly visualization of the sonication volume, targeted structures, and the calculated fields of ultrasound energy and the corresponding thermal maps. Visualization devices differ in image quality, form factor, and ease-of-use.

In summary, described contribution relies on the combined use of AI technologies with AR devices for the optimization of parameters for the application, from outside the body, of focused ultrasound beams to achieve preventive cleaning of deposits adhered to the inner surfaces of shunt systems (valves, catheters) implanted in patients with various pathologies.

ACKNOWLEDGMENTS

This project has been carried out in collaboration with FISEVI and University Hospital V. Rocio (Seville, Spain) and with the support of the following researchers: Manuel A. Perales-Esteve, Francisco J. Muñoz-Gonzalez, Desiree Requena-Lancharro, Isabel Fernández-Lizaranzu, Pedro Gil-Gamboa, Maria Jose Mayorga-Buiza and Mónica Rivero-Garvía.

WORKS CITED

ATTRACT Program. (2019). https://attract-eu.com/

Baraas, R. C., Imai, F., Yöntem, A.Ö. & Hardeberg, J. Y. (2021). Visual perception in AR/VR. *Optics & Photonics News* (April 2021), 34–41.

FUSCLEAN Project. (2020). https://phase1.attract-eu.com/showroom/project/combined-optical-imaging-and-ultrasound-focusing-for-hand-held-non-invasive-cleaning-of-implanted-cerebrospinal-fluid-shunting-devices-in-patients-of-hydrocephalus-initial-design-and-proof-of-concep/

Gómez González, E., & Gómez, E. (2020). *Artificial Intelligence in medicine and healthcare: Applications, availability and societal impact.* JRC Science for Policy Report (EUR 30197 EN), European Commission. https://doi.org/10.2760/047666

Nepelski, D. (Ed.). (2021). *How can Europe become a global leader in AI in health?* European Commission. https://knowledge4policy.ec.europa.eu/file/how-can-europe-become-global-leader-ai-health_en.

1.2 ASSURING THE QUALITY AND SECURITY OF MEDICAL ROBOTICS PROCESS AUTOMATION

María-José Escalona and José Navarro-Pando

ABSTRACT

In recent years, the application of new technologies to the healthcare environment is a common practice. However, the COVID-19 pandemic has shown us that this application is a critical necessity for society. The application of disruptive techniques such as Artificial Intelligence or Machine Learning in the healthcare environment is something necessary but not sufficient. It is necessary to take another qualitative leap. This paper presents a reflection on the use of emerging technologies, such as blockchain or the robotization of software processes to further improve healthcare processes.

INTRODUCTION

In recent years, many lines of research and innovation have been developed, aimed at improving the health environment through ICT (Information and Communication Technologies) and, very specifically, through the development of advanced software. In this sense, the use of technologies such as Artificial Intelligence or Machine Learning, the intelligent treatment of large masses of data (Big Data), or the involvement of complex software processes has had many fields of research and application. However, frequently the development team forgets that the development of this software has to be accompanied by important quality principles. Thus, guaranteeing aspects such as information security, reliability, or robustness of the software is an activity as basic and necessary as the need to produce the software itself. This work is based on presenting how a clinical process can be benefited if early principles of quality assurance are applied. In the work, it is presented how smart contracts and blockchain technology can be applied for the early assurance of the security and traceability of information. We also propose the use of RPA (Robotic Process Automation) to increase the automation of these processes and the reduction of errors in order to increase reliability and robustness.

STATE OF THE ART

The healthcare processes are the basis of many of the medical protocols that are developed and applied. In a very concrete way, the clinical guidelines show the steps that healthcare teams must follow to carry out a specific disease. These processes, guides, or protocols are usually represented by business processes that, very frequently, are implemented in software solutions. The development of these software tools has helped healthcare teams to apply medical protocols. Currently, Artificial Intelligence is being applied in software solutions of healthcare processes not only for helping healthcare teams. They are also suitable for supporting clinical decisions or even learning from their actions with Machine Learning protocols.

However, reality asks us for even more. All software, whether sanitary or not, must comply with quality principles: be safe, be robust, and be effective and efficient, among others. These principles must be guaranteed from their conception and, in the case of healthcare processes software solutions, given the scope of application, these quality principles acquire a critical character. Software must be conceived in this way and software development teams must hardly work on guaranteeing this quality from the beginning of software production.

On the other hand, this early quality assurance cannot mean an increase in costs, time, or resources, neither in the development of the software itself nor in its subsequent use in the healthcare environment.

In this work, we focus on three concepts: early safety, robustness, and reliability. There are few papers that early address these non-functional requirements in healthcare software generally and in healthcare process software solutions in a concrete way.

In the study preceded by Sánchez-Gómez (2020) a situation analysis is carried out in which proposals are evaluated to analyze aspects of safety, traceability, and reliability in technological solutions applied to laboratory environments, specifically human reproduction, seeing that there is a gap in this aspect.

Other disciplines, such as banking, have successfully applied blockchain-based environments to guarantee aspects of information security in transactions. In these environments, a series of smart contracts are defined to ensure that the information is transferred in an adequate mode. The definition of smart contracts in healthcare processes and the effective testing of them in early stages can be a very positive aspect for the assurance of information (Martins et al., 2019).

In the industrial environment, software process robotization has also been used for years to increase automation and, therefore, reduce errors, increase productivity and increase the reliability of results. The application of these principles can also be a key point in the environment of healthcare processes software solutions (Osman & Ghiran, 2019).

With all this, what can be seen is that specific aspects such as smart contracts or process robotization offer a further step in the context of health processes, providing them even more with greater efficiency and, therefore, better service for society.

CONTRIBUTION

With the arrival of the COVID-19 pandemic, the Ministry of Health issued a technical document to indicate the emergency management process for COVID-19 (Ministerio de Sanidad, 2020). Concretely, this document indicates how to deal with patients in order to identify, isolate and inform who may be affected by the disease as soon as possible. It obviously reduces the risks both for them, as well as for other patients, and for the sanitary staff themselves. This protocol can be modeled as a healthy process in which the sanitary staff interacts with the patient to, based on a series of rules, make the decision on how to act.

This protocol is a simple and very attractive example for the purposes of our research. Therefore, in collaboration with the Inebir clinic and the G7 Innovation company, a proof of concept was carried out. It allowed us to validate the use of smart contracts and RPA for the early assurance of non-functional requirements.

The proof of concept was carried out in different phases. The first thing that was done was to model the process itself defined by the Ministry. To do this, the entire team worked together applying the NDT 4.0 development methodology (Escalona & Aragón, 2008). Once the process was defined, the information transfer that was carried out was studied and modeled using the models that this methodology has for this. With all this, smart contracts were early defined that allowed the information provided by the patient to be safely exchanged through the triage protocol.

After that, it was analyzed where the process could be robotized. We concluded that the process itself was fully automatable as long as we could simulate the figure of the interviewer, in this specific case, the sanitary who was performing the triage.

In order to carry out all the proof of concept, we created a software application (available in desktop and APP format) based on a chatbot. Using this application, the patient was being interviewed by a virtual doctor, called Dr. Jones, who does the triage to the patient. The patient was answering the questions interacting with Dr. Jones. At the end of the process, the sanitary responsible for triage received a report showing the results: positive triage, if the patient was not a potential COVID patient, or negative triage otherwise.

With our proof of concept, we can conclude several things:

1. Our solution cannot make any decisions, what it actually does is apply a process in an automated way and give the result to the sanitary that, based on the result, will issue the final assessment.

2. The proof of concept, in addition to being reliable and demonstrating that our starting hypotheses regarding the use of smart contracts and RPA in sanitary processes are adequate, shows that solutions like this can have important benefits for society. Specifically, with our solution we have been able to see that:

   a)   This kind of software solution can help unblock the healthcare environment. If a patient performs the COVID triage with a software robot, the sanitary will not have to spend time in performing the triage, they will simply analyze the results, leaving the robot to do the more mechanical work.

   b)   Solutions of this type can help, in the specific case of COVID-19, to reduce infections. The patient can perform triage at home, even before arriving at the health center. If they turn out to be a possible COVID patient, their contact with other patients and with the sanitary staff themselves will be less and, therefore, the isolation will be more effective.

3. The proof of concept shows that, although the application of Artificial Intelligence is necessary, it must be applied with basic principles of software quality, and with our experience, we have shown that this does not imply making the solution or the development process more expensive.

Regardless of our solution, to carry out any technological application, it is necessary that the technical and functional teams, in this case, the sanitary ones, work collaboratively and jointly applying mechanisms, such as NDT 4.0, for the work to be effective and with results.

ACKNOWLEDGMENTS

This project has been carried out in collaboration with Inebir Technology S.L, and within the Web Engineering and Early Testing research group.

WORKS CITED

Escalona, M. J., & Aragón, G. (2008). NDT. A model-driven approach for web requirements. *IEEE Transactions on Software Engineering, 34*(3), 377–390.

Martins, G. D., Gonçalves, R. F., & Petroni, B. C. (2019). Blockchain in manufacturing revolution based on machine to machine transaction: A systematic review. *Brazilian Journal of Operations & Production Management, 16*(2), 294–302.

Ministerio de Sanidad. (2020). *Manejo en Urgencia del Covid 19*. Disponible en: https://www.mscbs.gob.es/en/profesionales/saludPublica/ccayes/alertasActual/nCov/documentos/Manejo_urgencias_pacientes_con_COVID-19.pdf

Osman, C. C., & Ghiran, A. M. (2019). When industry 4.0 meets process mining. *Procedia Computer Science, 159*, 2130–2136.

Sánchez-Gómez, N., Morales-Trujillo, L., Gutiérrez, J.J., & Torres, J. (2020). The importance of testing in the early stages of smart contract development life cycle. *Journal of Web Engineering, 19*(2), 215–242. https://doi.org/10.13052/jwe1540-9589.1925

1.3 MACHINE LEARNING AND SINGLE-CELL TRANSCRIPTOMIC ANALYSIS IN THE BRAIN

Marina Valenzuela-Villatoro and Rafael Fernández-Chacón

ABSTRACT

A major social challenge of modern societies is to prevent and treat neuro-degenerative diseases suffered by the constantly increasing elderly population. We are investigating the changes in gene expression in single neurons in response to nerve terminal dysfunction in the brain of genetically modified mice. Machine Learning approaches imported from other disciplines are key for the bioinformatic analysis of data sets comprising thousands of cells expressing thousands of genes each. Expression changes in specific genes might be essential to maintain neuronal homeostasis and, therefore, potentially useful to be targeted in the context of therapeutic strategies.

INTRODUCTION

Animals and plants function thanks to the coordinated interaction of multiple different cell types. The cell is the basic functional unit in biology. The specialized features of every cell depend on the selected repertoire of thousands of genes that every cell expresses at any time. Genetic information is stored in the genomic DNA. When a gene is expressed, the genomic DNA is transcribed into the so-called messenger RNA (mRNA). The knowledge of the dynamics of genetic expression, which determines which genes are expressed and which are not, is of fundamental importance to understanding normal cell function and dysfunctions linked to diseases. Recent revolutionary methodological advances have allowed separating individual cells from multicellular organisms; and using high-throughput sequencing techniques (e.g., RNA-seq), it is nowadays possible to identify virtually all the mRNA molecules in a single cell (single-cell RNA-seq).

This allows classifying cell types and detecting deviations between healthy and sick cells with unprecedented precision. The research carried out by the Tabula Muris international consortium exemplifies the advances in this field. This consortium has generated an atlas of cell types by analyzing the single-cell transcriptome (the whole set of mRNA molecules present in one single cell) in more than 100,000 cells from 20 different organs and tissues of a key model organism in biomedical research, the laboratory mouse (*Mus musculus*) (Schaum et al., 2018). Regarding brain analysis, these approaches are being extremely powerful to define the different neuronal types in the mouse brain (Zeisel et al., 2015) or to identify gene expression changes in the brain of Alzheimer's disease patients (Mathys et al., 2019). Bioinformatics based on Machine Learning approaches is key for the analysis of the myriad of data generated in these experiments.

STATE OF THE ART

Bioinformatics and computational methods applied to the analysis of single-cell RNA-seq (scRNA-seq) are being developed in part thanks to the Machine Learning approaches that previously existed in other fields (Raimundo et al., 2021; Butler et al., 2018). Standard analysis of scRNA-seq data sets might involve the deep analysis of hundreds of thousands of cells expressing each a selection of several thousands of genes out of a repertoire of more than 20,000 genes. This analysis normally follows a number of sequential steps including at least (1) the reading of raw data, (2) the elaboration of a count matrix, (3) the transformation of data into a low-dimensional space, (4) the clustering and taxonomic classification of cells into different cell types, (5) the inference of cell trajectories to understand the transformation of one type of cell into a different one (e.g., during the development or under any biological situation that implies progressive changes of gene expression), and (6) the discovery of differential gene expression patterns associated with a particular situation such as a pharmacological treatment or a particular genetic modification to model a disease. Machine Learning approaches are being determinant in an enormously growing list of tools devoted to scRNA-seq analysis. The website scrna-tools .org currently tracks more than 900 tools classified in 30 different categories. Among the analytic steps stated above, the identification of subpopulations of the cells present across multiple data sets is a major challenge. A successful analytical strategy to integrate scRNA-seq data sets, based on common sources of variation, has been implemented in the R toolkit Seurat V.2 developed by the laboratory of Rahul Satija[1] (Butler et al., 2018). This approach uses Machine Learning to extract shared gene correlation structures conserved between all the different data sets to be analyzed. Briefly, the procedure involves first the detection of highly variable genes, which are genes which expression varies significantly within cells and therefore show a high standard deviation. Based on those genes, it is possible to obtain a dimensionality reduction consisting of a simplification of the repertoire of genes containing only genes that are meant to describe specific cell types. Dimensionality reduction is reached using canonical correlation analysis (CCA) to generate canonical vectors or subspaces. Each vector is formed by a specific group of genes, a so-called metagene, that describes a specific group of cells. The alignment of canonical vectors through a dynamic time warping algorithm yields a shared low-dimensional space in which clusters of similar cells populations are easily visualized in 2D plots based on t-distributed Stochastic Neighbor Embedding (t-SNE) or Uniform Manifold Approximation and Projection (UMAP) (Butler et al., 2018). State-of-the-art approaches are moving beyond the analysis of single-cell transcriptomics and toward the comprehensive integration with other single-cell data such as epigenomic, proteomic, and spatially resolved single-cell data. This has been recently achieved and implemented into Seurat V.3 (Stuart et al., 2019).

CONTRIBUTION

Neurodegenerative diseases are an increasingly growing public health problem; however, the molecular mechanisms underlying synaptic and neuronal degeneration are poorly understood. Cysteine String Proteinα/DNAJC5 (CSPα/DNAJC5) is a synaptic co-chaperone related to neurodegeneration in humans (Valenzuela-Villatoro et al., 2018). Adult-onset autosomal dominant neuronal ceroid lipofuscinosis is caused by mutations in the gene DNAJC5. Mice lacking CSPα/DNAJC5 present activity-dependent synaptic degeneration, detected especially in the fast-spiking GABAergic interneurons that express parvalbumin (PV). Since these mice die early, it is difficult to follow the time course of the synaptic and cellular dysfunction of these neurons. We have generated a conditional knock-out mouse line lacking CSPα/DNAJC5 specifically in parvalbumin-positive GABAergic interneurons (Valenzuela-Villatoro, 2019). These mice develop a progressive neurological phenotype and synaptic dysfunction. In order to get insight into the mechanisms of gene expression associated with synaptic dysfunction in PV interneurons lacking CSPα/DNAJC5, we have carried out the analysis of single-cell transcriptomes of cortical PV+ interneurons in control and mutant mice. In collaboration with Dr. Ana Belén Muñoz-Manchado and Dr. Jens Hjerling-Leffler (Karolinska Institute), we have combined fluorescence-activated cell sorting (FACS), the

STRT-seq-2i method, and the WaferGen 9600-well platform for single-cell RNA sequencing. Next, we have conducted computational analysis using first the R toolkit Seurat V.2 and then Seurat V.3 (in collaboration with Dr. María Eugenia Saez-Goñi and Dr. Antonio González-Pérez, Centro Andaluz de Estudios Bioinformáticos (CAEBI)). We have integrated scRNA-seq data from control and mutant mice and, based on common sources of variation, we have identified different populations of PV interneurons and carried out a downstream comparative analysis of gene expression.

The gene ontology (GO) study based on the differential gene expression analysis suggests genetic dysregulation of metabolism and synaptic function, among other processes (Valenzuela-Villatoro, 2019). The collaboration with CAEBI has allowed a deeper bioinformatic analysis of cell clustering and differential expression analysis, which additionally included analysis of cell trajectories and investigation of interconnected genes (co-expression networks). A number of relevant transcripts have been selected and their changes have been experimentally investigated using quantitative measurements of RNA expression in situ by RNA-scope.

ACKNOWLEDGMENTS

This project has been carried out in collaboration with Andalusian Center for Bioinformatics Studies (Centro Andaluz de Estudios Bioinformáticos, CAEBI) and with the support of the following researchers: Ana Belén Muñoz-Manchado, José A. Martínez-López, and Jens Hjerling-Leffler (Karolinska Institute, Stockholm, Sweden)

Note

1. https://satijalab.org/seurat

WORKS CITED

Butler, A., Hoffman, P., Smibert, P., Papalexi, E., & Satija, R. (2018). Integrating single-cell transcriptomic data across different conditions, technologies, and species. *Nature Biotechnology*, *36*(5), 411–420. https://doi.org/10.1038/nbt.4096

Mathys, H., Davila-Velderrain, J., Peng, Z., Gao, F., Mohammadi, S., Young, J. Z., Menon, M., He, L., Abdurrob, F., Jiang, X., Martorell, A. J., Ransohoff, R. M., Hafler, B. P., Bennett, D. A., Kellis, M., & Tsai, L. H. (2019). Single-cell transcriptomic analysis of Alzheimer's disease. *Nature*, *570*(7761), 332–337. https://doi.org/10.1038/s41586-019-1195-2

Raimundo, F., Papaxanthos, L., Vallot, C., & Vert, J.-P. (2021). Machine learning for single cell genomics data analysis. *BioRxiv*, 2021.02.04.429763. https://doi.org/10.1101/2021.02.04.429763

Schaum, N., Karkanias, J., Neff, N. F., May, A. P., Quake, S. R., Wyss-Coray, T., Darmanis, S., Batson, J., Botvinnik, O., Chen, M. B., Chen, S., Green, F., Jones, R. C., Maynard, A., Penland, L., Pisco, A. O., Sit, R. V., Stanley, G. M., Webber, J. T., … Weissman, I. L. (2018). Single-cell transcriptomics of 20 mouse organs creates a Tabula Muris. *Nature*, *562*(7727), 367–372. https://doi.org/10.1038/s41586-018-0590-4

Stuart, T., Butler, A., Hoffman, P., Hafemeister, C., Papalexi, E., Mauck, W. M., Hao, Y., Stoeckius, M., Smibert, P., & Satija, R. (2019). Comprehensive integration of single-cell data. *Cell*, *177*(7), 1888–1902.e21. https://doi.org/10.1016/j.cell.2019.05.031

Valenzuela-Villatoro, M. (2019). *Single-cell transcriptomic and functional characterization of cortical parvalbumin interneurons in a novel conditional knock-out mouse lacking CSPα/DNAJC5* [Ph.D. Thesis, Universidad de Sevilla]. https://idus.us.es/handle/11441/88149

Valenzuela-Villatoro, M., García-Junco-Clemente, P., Nieto-González, J. L., & Fernández-Chacón, R. (2018). Presynaptic neurodegeneration: CSP-α/DNAJC5 at the synaptic vesicle cycle and beyond. *Current Opinion in Physiology*, *4*, 65–69. https://doi.org/10.1016/j.cophys.2018.06.001

Zeisel, A., Munoz-Manchado, A. B., Codeluppi, S., Lonnerberg, P., La Manno, G., Jureus, A., Marques, S., Munguba, H., He, L., Betsholtz, C., Rolny, C., Castelo-Branco, G., Hjerling-Leffler, J., & Linnarsson, S. (2015). Cell types in the mouse cortex and hippocampus revealed by single-cell RNA-seq. *SCIENCE*, *347*(6226), 1138–1142. https://doi.org/10.1126/science.aaa1934

Energy Efficiency and Sustainable Construction

2

The importance of sustainability is key in the long-term industrial impact of Andalusia, and that is why it is detected as one of the strengths of the University of Seville research groups, reflected in their collaboration with the private sector and the impact they all make in the sector to achieve excellence. In recent years, energy efficiency and sustainability have been promoted, enhancing Artificial Intelligence techniques, such as Machine Learning, Big Data, the Internet of Things (IoT), and Business Intelligence.

Following these objectives, an IoT application for the gas sector is presented in this chapter, setting out a low-cost electronic sensor technology, hardware design, and the design of the pipeline, in order to have an intelligent measurement system of individualized gas energy consumption in homes and industrial companies, maximizing the battery life of these sensors.

Another application covered in this chapter is related to the design of new wireless sensors for intelligent systems in the construction and transport sector that guarantee sustainable operations. To do this, a unified information storage system that applies Artificial Intelligence techniques is considered, by increasing its efficiency and productivity.

Finally, quantitative improvements in intelligent predictive maintenance applied to rail transport are put forward. A condition-based maintenance process is designed, which facilitates the integration of historical data, translating it into the aid of dynamic maintenance decision-making, particularized in the train axes.

DOI: 10.1201/9781003276609-2

2.1 IOT FOR ENERGY-EFFICIENT GAS METERS

José Ramón García Oya and Ramón González Carvajal

ABSTRACT

In this paper, a complete design of a gas flow sensor based on the measurement of the time of flight (ToF) of an ultrasonic signal is presented. After the ultrasound-based technique has been selected, the sensor design flow has consisted of the following main stages: mechanical pipe design, hardware selection and design, ToF detection algorithm selection and implementation, and piezoelectric transducer selection. As a result of this design, an ultrasonic-based gas flow sensor is presented, which completely fulfills the accuracy requirements given by the standard EN14236, improving the current commercial gas flow sensors specifications regarding costs, ultrasonic signal-to-noise-ratio, and power consumption, which is crucial to implement smart gas meters able to autonomously operate as IoT devices by extending their battery life.

INTRODUCTION

Nowadays, smart meters are becoming an essential instrument for improving energy efficiency. This paper proposes a new smart gas flow sensor based on ultrasonic communication for energy-efficient applications in gas distribution networks, compatible with battery durations of more than 15 years, the ability to detect the type of gas to be measured, and equipped with Artificial Intelligence (AI) through IoT technology, to provide services related to the digital economy and also with the ability to adapt the total consumption of the system based on the flow rate that it measures. The sensor will be integrated into a smart meter and will be tested under real conditions.

The complete system will be used for the integrated, efficient, and sustainable management of gas distribution networks, allowing public or private operators access to detailed information of the state of the network, from the generation of the resources to its delivery to the client (domestic or industrial). The information captured by the sensor will generate value through AI, allowing features such as demand forecast, exact knowledge of consumption patterns, network load status, diagnosis of the infrastructure, and maintenance forecasts. In addition, it will promote direct communication with the client, implying a responsible use of energy. The system will use IoT technology to provide connectivity and intelligence, and it will be industrialized and tested in a real operating environment, in collaboration with the company WOODSWALLOW.

STATE OF THE ART

The selected technology to measure the gas flow is based on ultrasonic communication, whose operating principle is the measurement of the time of flight (ToF) of the ultrasonic wave, being possible to measure the flow velocity without dependency with the medium and also measuring the velocity of the sound in the medium without dependency with the flow velocity (Baker, 2016). This feature will be used as the basis for the calibration procedure implemented by AI in order to correct the flow measurement errors due to temperature variations.

Some benefits of the ultrasonic-based sensors are the following:

- Accuracy: it can be calibrated to <0.1%
- Non-intrusive: minimal pressure drops and non-obstruction of the flow
- Bidirectional: measurement of volumes in both directions
- The composition of the gas inside of the meter could be unknown: the measurement is independent of the theoretical value of the speed of sound
- Gas properties: Additional information about gas properties (such as sound velocity profile), which might be used for density and calorific value determination
- Possibilities for fast time response: measurement of pulsating flow
- Installation: easy to install
- Low maintenance: no moving parts means reduced maintenance tasks, compared to mechanical-based sensors
- Potentials of remote operation in order to implement the AI operations described in the Introduction section
- Ability to self-diagnose the meter's health: it is possible to validate a proper operation by electronic diagnosis from parameters such as sound velocity and signal level

However, there are several challenges related to ultrasonic-based sensors that should be addressed:

- Sensitivity to ultrasonic noise and parasitic ultrasonic signals
- Relatively uncertain sensitivity to installation conditions (bends, pipe roughness, flow conditioners, etc.)
- Sensitivity to turbulences and asymmetrical flow profiles across the path tube and at the transducers cavities
- Influence of the propagation phase difference between both ultrasonic paths
- Insertion loss at the transducer's clamp-on material

Once the ultrasonic-based technology has been selected, the design flow has been conducted to achieve several innovations regarding the state of the art of the current commercial ultrasonic-based sensors, such as:

- Signal-to-noise ratio (SNR) enhancement by digital signal post-processing
- Power consumption minimization
- Ability to work with different gas compositions

CONTRIBUTION

The proposed gas flow sensor based on ultrasonic communication involves different design stages:

a) *Mechanical design of the flow path tube*

In this stage, different ultrasonic paths configurations have been tested: Z-configuration (based on a direct way between the transducers), V-configuration (based on one reflection of the ultrasonic wave over the pipe floor), and W-configuration (based on two reflections of the ultrasonic wave over the pipe floor). For each configuration, different form-factor alternatives have been tested: incidence angle, ultrasonic path length, and cross section. Additionally, different materials (such as PLA, SLA, and ABS by using a 3D printer) and different lamination and collimating elements have been experimentally characterized.

After this study, a V-configuration results in the best option regarding:

- Sensitivity: although a Z-configuration usually provides a more robust transmission (higher SNR), it has a lower sensitivity for these applications, because there is a lower contact between the ultrasonic signal and the gas.

- Simplicity: moreover, the reflective V-configuration allows a simpler assembly, because both transducers are placed at the same side of the pipe.

The best option regarding the accuracy of the selected V-configuration has the following mechanical dimensions: an incidence angle of 65°, a path length of 55.9 mm, and a rectangular cross section of 20.3×9.5 mm^2. By employing a rectangular cross section, it is possible to allow a flow measurement under the same condition from the upstream side to the downstream side of the ultrasonic wave propagation path. In addition, three planar separation plates have been placed into the pipe to make the flow velocity distribution more uniform. Finally, the collimation elements are implemented by using aperture holes sealing molded with the transducer-fixing casing, in order to stabilize the flow between the ultrasonic transducers, improving the ultrasonic reception level, and thereby increasing the measurement precision and reducing the driving input for the ultrasonic transducers.

b) *ToF detection algorithm selection*

A correlation-based method by using an analog-to-digital converter (ADC) has been selected instead of a zero-crossing-based method by using a time-to-digital converter (TDC). Using the correlation technique, the whole waveform is captured, and the digital post-processing is performed to derive the differential ToF, whereas that the zero-crossing technique is based on the detection of the zero signal levels without capturing the whole waveform. By using the correlation technique, it is possible to achieve better accuracy performance, since the correlation acts as a digital filter reducing the noise and other interferers. Also, it is more robust to amplitude signal variations, at the expense of higher power consumption per flow measurement.

However, this high-accuracy correlation-based method allows to perform a lower number of measurements to achieve the uncertain requirements given by the standard EN14236; so from a preliminary power consumption estimation (19.2 µA·s per flow measurement), it will be possible to perform only 3–4 measurements each 2 seconds (measurement time required by the standard), enough to comply with the norm specifications for reducing the energy consumption.

Additionally, the sensor parameters have been optimized (at accuracy and power consumption levels) depending on the number of transmitted pulses, the sampling frequency, the use of a single-tone or multi-tone transmission mode, and the selection of the algorithm for the absolute ToF (upstream and downstream) measurement. This algorithm is based on the Hilbert transform, in order to implement the auto-correlation of the received signal (Hanus, 2015).

The accurate measurement of these absolute times is essential to detect temperature variations in order to calibrate the flow measurement accordingly and also to avoid the use of an external temperature sensor, which would increase the cost and the power consumption.

c) *Hardware selection*

The electronic circuitry integrated into the sensor is based on COTS (commercial off-the-shelf), specifically on the Texas Instruments MSP430FR6043 component. This ultrasonic front end has been selected in order to minimize the power consumption and due to its capabilities to implement the selected correlation-based method by using an internal microprocessor. Other studied ultrasonic front ends have been the Maxim MAX35104, the AMS TDC-GP30, the Texas Instruments TDC1000 + TDC7200, and the Maxim MAX35101.

d) *Piezoelectric transducer selection*

An experimental comparison between different piezoelectric manufacturers (Ceramtec, Jiakang, etc.) and between different nominal ultrasonic frequencies (200–400–500 kHz) has been performed. These different transducers have been evaluated in terms of sensitivity, uncertain flow measurement, zero flow drift performance, and flow measurement deviation between pairs in a climatic chamber.

A nominal frequency of 200 kHz has been selected because it presents better properties regarding sensitivity and gas attenuation (Ejakov et al., 2003), relaxing the requirements of the pre-amplification stage and reducing the hardware costs and the energy consumption. Finally, the transducer Jiakang 200KHz has been selected because of its zero flow drift performance (±0.6 ns for the tested pairs, i.e., <3% of the minimum differential ToF at 40 l/h required by the standard, in the range $-10°C/40°C$) and its flow measurement deviation between pairs for different temperatures and flow rates (with a maximum deviation of $\approx1\%$), assuming the pairs are calibrated at ambient temperature (23°C).

e) *Hardware design*

The printed circuit board (PCB) installed in the gas meter is based on the MSP430FR6043 ultrasonic front end, and it has been manufactured and experimentally characterized. Different versions of this hardware have been implemented in order to optimize the transducer impedance matching, and the filtering and the low-noise pre-amplification stages.

Finally, by the integration of all previously designed parts and by using an experimental setup based on several compressors and flow meters, the obtained accuracy results fulfill the standard EN14236, which requires an admitted error of (averaging six measurements) <3% in the range of 40–600 l/h and <1.5% in the range of 600–7200 l/h. Specifically, the proposed sensor presents a maximum error of 1.8% at 40 l/h, <1% in the range 80–600 l/h, and <0.5% in the range 600–7200 l/h.

An intelligent and auto-calibrated ultrasonic-based gas flow sensor has been presented in this paper, with an enhanced signal-to-noise ratio and minimal power consumption. All the sensor design flow stages (at mechanical, hardware, and firmware levels) have been developed and experimentally validated, obtaining an accuracy performance that fulfills the standard requirements, with the purpose of integrating into smart gas flow meters, providing new functionalities and features, such as extended battery life, efficient control of the energy consumption, the ability to adapt to all markets and types of gas, the prediction of demand, and the maintenance of the infrastructure. The impact of the proposed IoT gas meter will open new businesses in the digital economy, such as new forms of relationship between the client, the infrastructure operator, and the regulator.

ACKNOWLEDGMENTS

This project has been carried out in collaboration with WOODSWALLOW S.L. and with the support of the researcher Fernando Muñoz Chavero.

WORKS CITED

Baker, R. C. (2016). *Flow measurement handbook. Industrial designs operating principles, performance and application*. Cambridge University Press.

Ejakov, S. G., Phillips, S., Dain, Y., Lueptow, R. M., & Visser, J. H. (2003). Acoustic attenuation in gas mixtures with nitrogen: Experimental data and calculations. *The Journal of the Acoustical Society of America, 113*(4), 1871–1879.

Hanus, R. (2015). Application of the Hilbert Transform to measurements of liquid–gas flow using gamma ray densitometry. *International Journal of Multiphase Flow, 72*(2015), 210–217.

2.2 WIRELESS SENSOR NETWORK FOR SUSTAINABLE CONSTRUCTION AND INTELLIGENT TRANSPORT

Pedro Blanco Carmona and Antonio Jesús Torralba Silgado

ABSTRACT

This paper presents the design of a new sensor node for intelligent systems, combining the needs of the IoT (wireless connectivity, low consumption, simplicity and reduced cost), with limited local processing capacity, to respond to the needs of new edge- and fog- computing paradigms that appear as a consequence of the increase in the complexity of the systems and in the application of Artificial Intelligence techniques in the fields of sustainable construction and intelligent transport systems.

INTRODUCTION

The massive deployment of wirelessly interconnected sensing elements, as well as the ubiquitous presence of sensors in electronic devices, household appliances, automobiles, mobiles, and laptops, and the need to integrate this information with that coming from computers and people has led to the so-called IoT. In parallel, the development of systems based on Artificial Intelligence makes it possible to extract knowledge from the information available for decision-making and to develop systems that learn from experience, and even react to unforeseen stimuli, imitating human behavior. The massive availability of data from the IoT entails new paradigms of action that require distributed computing, giving rise to concepts such as Cloud Computing on massive data sets using Big Data techniques.

STATE OF THE ART

This paradigm is changing recently as a consequence of (1) the need to reduce the amount of relevant information that is exchanged with the cloud to accelerate the processing of information, (2) the need reduce the communications of the sensor elements since these are the most demanding in consumption and limit the useful life of autonomous devices, and (3) the possibility of increasing the local processing capacity in the nodes, as a consequence of the appearance of microprocessors with high processing capacity, ultra-low power consumption, and sufficient memory size. As a consequence, cloud processing is combined with new paradigms such as edge- or fog-computing (Bibri, 2018). All this leads us to the need to develop new sensor nodes with intermediate capacity (Zantalis et al., 2019), which is the objective of this project.

Azvi, the collaborating company on the project, is an international reference in sustainable construction and transport infrastructures, having collaborated with the research group in numerous R&D projects (García-Castellano et al., 2019), (Torralba et al., 2021). In these fields of application, massive use of the IoT is necessary to face the new challenges that arise in terms of monitoring and maintaining critical infrastructure for transport and sustainable construction. Throughout this collaboration, both entities have verified the need to develop this type of intermediate sensor node while maintaining the flexibility to adapt to the different sensors that are currently used, keeping in mind a future adaptation of the sensor node to the IoT standards of the fifth generation of the mobile technology.

CONTRIBUTION

This article presents the development of a sensor node for the IoT, with low cost, small size, and moderate processing capacity, especially oriented to monitoring in sustainable construction and transportation. It is based on LoraWan technology, which includes interfaces with the most common sensors required by the collaborating company. As part of the technology transfer, work is also done on the industrialization of the node and the testing of the industrialized and pre-certified node in an operational environment.

The collaborating company has defined the specifications of the wireless nodes in relation to their applications of interest, focusing on an open architecture that allows their adaptation to the different sensors. These specifications have served as starting data to define the architecture of the sensor node. It is relevant to indicate here that different sensors have been evaluated to verify the acquisition and processing capacity of the mode with respect to the data coming for different sensors, and several evaluation PCBs have been built for this purpose. Special attention has also been paid to the integration of the LoRaWan communication module. However, it has been taken into account that the architecture is sufficiently modular and efficient so that the communications module can be replaced by a technology compatible with 4G and 5G such as NBIoT.

The sensor node is based on an ultra-low-power ARM Cortex-M3 microcontroller with 512 Kbytes of flash memory, 32 MHz of CPU, and a memory protection unit. It admits power from 1.65 to 3.6 V, and so it can be operated by batteries. This microcontroller (STM32L152RE) is in charge of running the operating system in real time (free Real Time Operating System [RTOS]) and has a driver compatible with RTOS for each of the sensors selected for each type of node. It also has a driver to control a LoRaWan modem by AT commands. The project also designs the concentrator node, based on the same architecture as the sensor node, but with the ability to communicate with a gateway, based on AT commands, to be accessible from the LoRaWan backbone.

Once the sensor nodes have been manufactured and tested, a test network has been set up, on which different intelligent monitoring and control algorithms will be tested, in order to verify the system's capabilities for intelligent acquisition and fog-computing functions. For this, intelligent road traffic control systems will be used as an example, which use Artificial Intelligence to manage rail traffic in singular environments (Torralba et al, 2020) and to control traffic in unattended intersections in rural and peri-urban environments, which include a neuro-fuzzy system for remote learning and updating decision rules and scenario classification (Torralba et al, 2021). In this phase, the collaborating company will play a very significant role, verifying that the set specifications are met.

Finally, the final tests will be carried out, and the node will be redesigned for its industrialization, ending with a level of technological maturity of TRL8.

ACKNOWLEDGMENTS

This project has been carried out in collaboration with AZVI and within the Electronic Engineering research group.

WORKS CITED

Bibri, S. E. (2018). The IoT for smart sustainable cities of the future: An analytical framework T for sensor-based big data applications for environmental sustainability. *Sustainable Cities and Society, 38*, 230–253. https://doi.org/10.1016/j.scs.2017.12.034

García-Castellano, M., González-Romo, J. M., Gómez-Galán, J. A., García-Martín, J. P., Torralba, A., & Pérez-Mira, V. (2019). ITERL: A wireless adaptive system for efficient road lighting. *Sensors, 19*(23), 5101. https://doi.org/10.3390/s19235101

Torralba, A., García-Martín, J. P., González-Romo, J. M., García-Castellano, M., Peral-López, J., & Pérez-Mira, V. (2021). AISCS: Autonomous intelligent sign control system using wireless communication and LED signs for rural and suburban roads. *IEEE Intelligent Transportation Systems Magazine*. https://doi.org/10.1109/MITS.2021.3049375

Torralba, A., García-Castellano, M., Hernández-González, M., García-Martín, J. P., Pérez-Mira, V., Fernández-Sanzo, R., Jácome-Moreno, A., & Gutiérrez-Rumbao, F. J. (2020). Smart railway operation aid system for facilities with low-safety requirements. *IEEE Intelligent Transportation Systems Magazine*. https://doi.org/10.1109/MITS.2019.2962148

Zantalis, F., Koulouras, G., Karabetsos, S., & Kandris, D. (2019). A review of machine learning and IoT in smart transportation. *Future Internet, 11*, 94. https://doi.org/10.3390/fi11040094

2.3 INTRODUCTION TO DYNAMIC MAINTENANCE SCHEDULING BASED ON THE ADVANCED USE OF PREDICTIVE ANALYTICAL TECHNIQUES: APPLICATION TO THE USE CASE OF INTELLIGENT MAINTENANCE OF HIGH-SPEED TRAINS

Adolfo Crespo Márquez and Antonio De la Fuente Carmona

ABSTRACT

Nowadays the industrial grow of intelligent maintenance not only demands predictive analytics techniques but new methods that allow translating them into dynamic maintenance decision-making. This paper presents an application of predictive analytics for intelligent maintenance of train axle bearings. It is included within the design of a complete Condition Based Maintenance general process that facilitates the integration of this data and results at information systems level and the human understanding of the information provided by these solutions in order to provide simple interconnection with the maintenance actions scheduling processes.

INTRODUCTION

The unprecedented advance in predictive maintenance is one of the pillars of Industry 4.0. This has led to an increase in the use of a wide range of sensors with the aim of achieving real-time monitoring of the variables that allow systems' control. The development of predictive analytics makes the usefulness of this data more concrete through the generation and use of algorithms for the detection, diagnosis, and prediction of failures. Once a certain level of maturity has been reached in these algorithms, interest has shifted, in recent years, toward algorithms for intelligent decision-making for the dynamic maintenance-task-planning process.

This work introduces condition-based maintenance (CBM) plan for a train bearing failure modes based on real-time temperature monitoring. The CBM solution aims to reduce the risk of failure occurrence, extend the service life of the bearings, and optimize their maintenance in terms of costs, establishing the guidelines for decision-making in the planning of maintenance tasks, adapting the response times for the planning of interventions on the failure modes analyzed. A neural network model capable of detecting the anomalous behavior of the bearing temperatures associated with the failure mode under maintenance is proposed.

STATE OF THE ART

Within the rolling system, axle bearings are critical from the point of view of service quality and availability to the customer. Although the failure rate of the elements making up these systems is very low—with most of the elements' incidents registered with a rate of less than 5×10^{-3} failures per million kilometers (FPMK)—more than 50% of these elements are considered critical elements given the severity of their functional loss. Contributions are emerging including the development of advanced analytics-enabled frameworks to monitor the deterioration of these systems, predicting their rate of deterioration, and recommending their maintenance for optimal scheduling to maximize expected lifetime (Kumar and KP, 2019). Improvements in computing and communication technologies allow automated data collection on a large scale. Prediction models can then be conveniently trained and put into operation with real-time sensors data. Clearly, businesses' digital transformation fosters a new Operations & Maintenance (O&M) paradigm: merging the data collected from assets and sensors with big data analytics allows to monitor entire fleets down to individual components and to plan maintenance actions only when needed (Ferroni et al., 2018).

Literature of predictive analytics application to intelligent maintenance in trains axle bearings includes, among others, references to Support Vector Regression, which has been proposed for the estimation of the Remaining Useful Life (RUL) of damaged bearings based on vibration frequencies and with good results but high computational requirements (Fumeo et al., 2015); artificial neural networks (ANN) for predictive maintenance, through the use of vibration sensors, acoustic bearing detector, and optical detectors; most recent papers on bearings predictive maintenance are based on the study of vibration data together with other signals allowing to bearing anomalies. The need for vibration data is the most relevant limitation required to be able to apply some of the pre-existing methods to this paper's case study. In addition, computation times to generate a balanced solution, with high precision to provide rapid decision-making, is also a relevant constraint.

On the other hand, advanced CBM frameworks should integrate predictive analytics (PdA) or prognosis health management (PHM) techniques. In an advance and complete CBM process conception, data analysis allows detection, diagnosis, and prognosis, and these three results can be used as basic inputs for maintenance planning, the final goal of the CBM process (Guillen et al, 2016).

CONTRIBUTION

As described above in the text, this research solution is composed of combining two aspects or components that can be included in every intelligent maintenance proposal: CBM general process and predictive analytic solution.

a) *CBM general process*
   When continuously receiving new information from monitoring or maintenance activities, a dynamic process of maintenance decision-making starts. These two data inputs must be differentiated, since monitoring is continuous and usually requires predictive analytics, and preventive maintenance activities are discrete and do not require these techniques. In general terms, it is possible to identify four main steps:
   • Predictive analytics: Predictive analytics is required to transform the information that arrives from the IoT sensor network into useful information that can be used in the subsequent subprocess of event interpretation rules.
   • Rules of interpretation of events: Predictive descriptors or maintenance activities performed are received in the event interpretation rules subprocess. In the case of predictive descriptors, they are compared with thresholds that have been previously defined in the event design.
   • Rules for assigning risk levels to events: the objective of this subprocess is to evaluate the new risk level of the item. The input of this subprocess is an event that will be analyzed to define a new level of risk for the item. Once an event has occurred and the

corresponding new risk level has been assigned, we proceed to the last point of the methodology, risk treatment, and the main input is the new risk level associated with the failure modes analyzed.

- Maintenance decision-making/maintenance scheduling: the last step of the methodology is to determine the preventive maintenance actions to be carried out and, in general, the maintenance schedule over time. Once an event has occurred and the new risk level has been assigned, we try to mitigate or reduce the risk through various actions. These actions are the result of the decision-making subprocess.

b) *Predictive analytics solution*

This analysis focuses on axle bearings in a fleet of passenger trains. The train fleet considered has a total of 16 trains. In addition, axle bearings typically have several major maintenance activities in the shop, restoring their functionality, and reach a service life of 4 to 5 million km. The temperature of each bearing is continuously monitored by onboard train control monitoring systems (TCMS). It is possible to implement an analysis to detect the anomalies in the temperature of the bearings of the train rolling system. The implementation of the predictive analytics strategy generally consists of three steps:

- Select a predictive model: The algorithms used for the analysis are generalized linear models (GLM), ANN, decision trees (DT), random forest (RF), gradient-boosted trees (GBT), and support vector machines (SVM). The comparative study of the different models was carried out with the Rapidminer Studio tool. The algorithm selected for the implementation of the strategy has been the neural network, due to the best relationship between the correlation coefficient vs. total time (training + implementation), assuming a minimum required correlation factor of 0.96.

- Training a model for prediction by selecting the most suitable data set: the data preprocessing starts by joining the 16 databases generating 1 database for training and production (containing Ti bearing data sets from any axle j of any train). The database used for training has a total of 14,000,000 records and the one to be used for the validation study will have 6,800,000 records. The second step consists of cleaning the databases. The last step is to store the database of approximately 1gigabite of information, in the repository, for later use in the learning process.

- Test and validate the model: historical data of train operation, which have been contrasted with information related to incidents registered in the Computerized Maintenance Management System (CMMS), and subsequently validated during maintenance. Data sets with approximately 6–8 months of operation have been selected in order to be able to appreciate the behavior of the AE (absolute error in temperature prediction) in the three periods in which the bearing condition is divided.

The results obtained allow us to affirm that the predictive analytics solution can detect anomalous behavior in the bearing temperature when the AE exceeds 10°C, anticipating the failure in at least 30,000 km.

ACKNOWLEDGMENTS

This project is carried out in collaboration with SOLTEL (STL) and within the SIM (Intelligent Maintenance System) research group.

WORKS CITED

Ferroni, F., Klimmek, M., Aufderheide, H., Laia, J., Klingebiel, D., & Davidich, M. (2018). Data driven monitoring of rolling stock components. In: Y. Bi, S. Kapoor, & R. Bhatia (Eds.), *Proceedings of SAI intelligent systems conference (IntelliSys) 2016. Lecture notes in networks and systems, vol 15* (pp. 1003–1013). Springer. https://doi.org/10.1007/978-3-319-56994-9_68

Fumeo, E., Oneto, L., & Anguita, D. (2015). Condition based maintenance in railway transportation systems based on big data streaming analysis. *Procedia Computer Science, 53*, 437–446. https://doi.org/10.1016/j.procs.2015.07.321

Guillén, A. J., Crespo, A., Gómez, J. F., & Sanz, M. D. (2016). A framework for effective management of condition-based maintenance programs in the context of industrial development of E-maintenance strategies. *Computers in Industry, 82*, 170–185. https://doi.org/10.1016/j.compind.2016.07.003

Kumar, M., & KP, A. (2019). Rolling bearing damage. Recognition of damage and bearing inspection. https://doi.org/10.13140/RG.2.2.33971.17447

Digital Economy 3

The digitization of the economy has been a reality for a long time and new products and services that aim to make life easier for citizens and companies have been developed, automating processes or facilitating remote management. In the same way, this type of companies and services generally provide a multitude of data that can be processed to gain efficiency, improve the way of treating the customer, or offer new products.

In this chapter, we can find a paper about incidents and the use of the Internet of Things, in order to obtain information from sensors and use it to make intelligent decisions. The paper specializes in cybersecurity in Smart Cities, applied to lighting systems in cities, compiling historical data, and applying Artificial Intelligence techniques to detect potential future threats.

Another example presented below is based on the governance of information technology infrastructures by public administrations, so that citizens have transparent information about what really happens. New platforms are presented under blockchain technology that guarantees transparency and efficient behavior.

Likewise, the use of Artificial Intelligence and Machine Learning techniques for the classification of data sets in conversational systems is presented. In the generation of new conversational systems, large data sets are needed, so using intelligent methodologies that reduce the manual workload and guarantee the quality of the data will make them faster in terms of implementation and advanced when it comes to their use.

In addition, taking advantage of the amount of information on internet websites, the progress made with Artificial Intelligence techniques is presented to retrieve content from different web portals and verify the quality of the hosted data. All this is particularized for the example of the directory of companies and establishments in Andalusia.

DOI: 10.1201/9781003276609-3

3.1 DETECTING CYBER INCIDENTS IN IOT: A CASE STUDY IN SMART CITIES

Rafael Mª Estepa Alonso and Antonio Estepa Alonso

ABSTRACT

Intrusion detection systems (IDS) enable the detection of cybersecurity incidents. This work addresses the design of an IDS able to detect anomalous behavior in a real-world Smart Lighting application in the context of Smart Cities. The system design does not interfere with the operation (i.e., passive) and applies AI techniques to the data collected from sensors. Results show that our system is effective for the detection of a wide range of potential threats.

INTRODUCTION

Smart Lighting systems constitute an application of the Internet of Things (IoT) in Smart Cities. Cyberattacks on Smart Lighting systems can have a significant impact on the safety of citizens. As such, early detection plays a critical role in the mitigation of its impact.

This project suggests a security solution for detecting cyber incidents in the Smart Lighting system developed by Wellness Techgroup, which has been deployed in more than ten countries. This system comprises lights controllers that communicate with an IoT application server (CoAP/MQTT) every 300 seconds to send data or receive instructions from the system manager. The controller turns the lights on/off according to circumstances, such as sunset or sunrise time and luminosity. Lights themselves can also be configured to trigger events because of motion detection or low luminosity. The communication between the IoT server and the controller uses a deployment-dependent network such as carrier-based NB-IoT or LPWA (i.e., Lora or SigFox). Besides cyber threats, the system under design should detect operational anomalies such as fraudulent power drain. Finally, the system design should be guided by the principles of minimum cost and minimum computational resources.

STATE OF THE ART

Cybersecurity in the Internet of Things (IoT) is a research sub-field within the broader context of industrial control systems (ICSs). The reader can find in the study preceded by Cheminod (2012) an excellent state of the art in ICS and its main differences with the IT world. These include different protocol stacks, devices with constrained resources, and data traffic more predictable. Cyberattacks in ICS also exhibit particularities with respect to its counterpart in IT. Daily news show how ICS attacks are performed by state-sponsored groups, being more sophisticated and addressed to a specific target (advance persistent threat or ATP). On a few occasions, these attacks use new vulnerabilities (0-day attacks), which explains why defense systems need to use techniques for detecting both known attacks (e.g., signature-based detection) and unknown attacks (e.g., anomaly-based detection) (Knowles et al., 2015). But a prerequisite for detecting abnormal behavior is to know a normality profile on the field that depends on the particular ICS application domain. For this reason, commercial products and proposals found in the scientific literature are typically customized for a specific application domain (e.g., power grid, gas plants, water-treatment plants, IoT, etc.) (Bhamare et al., 2020). In this project, we want to achieve an intrusion detection system suited for Smart Lighting.

The main threats to be detected are:

- Attacks targeted to the IoT server. These can be carried out by compromising the lights controller (located in the carrier-operated NB-IoT network), which could lead to actions such as poisoning the temporal series database or performing denial-of-service (DoS) attacks.
- Credential theft. The stealing of admin credentials enables the system's malicious configuration, which could lead to undesired outcomes such as turning off lights at night.
- Operational issues. It would be desirable to detect incidents such as fraudulent electricity connections or malfunction of equipment.

Other system components (e.g., a management application) may be attacked. Still, we believe that existing technology from the IT world such as web application firewalls or conventional IDS (e.g., Snort, Suricata, etc.) suffices for a decent level of protection for these components. Indeed, our solution design should be compatible with these systems (which operate based on signatures or malicious patterns).

Finally, the solution design should be constrained by some requirements. We highlight the following: the use of Artificial Intelligence (AI) techniques for detecting abnormal system behavior (e.g., ATP or 0-day attacks), and the minimization of the consumption of resources, either computational, network bandwidth, or cost.

CONTRIBUTION

This project has enabled the collaboration between the university and the private sector (Wellness Techroup). The local company has provided its Smart Lighting product deployed in several worldwide locations and a real-life data set that includes three months of operation, which has been critical for training, testing, and validating the AI techniques in place.

We have designed an intrusion detection system made up of the following modules:

- B1: traffic-based anomaly detection module. This module takes a copy of the traffic received by the IoT server over a period (default is 5 minutes) and generates events that indicate abnormal behavior. To do this job, it creates a flow-based traffic matrix and characterizes several properties of each communicating pair. These properties are examined for anomalies via 25 indicators that use statistical techniques such as EWMA or decision trees. Some of the defined indicators were the number of destination ports reached from one source address, IP addresses from unexpected countries, or included in reputational databases. This module watches anomalies at the traffic level (OSI layers 1–4). As such, the range of potential detections includes scanning attacks, DoS attacks, unauthorized communications, abnormal traffic patterns, etc.
- B2: application-level anomaly detection module. This module takes the data received from controllers and generates events that can be indicative of a compromised IoT server (e.g., credential theft) or operational issues such as fraudulent electric connections, erroneous configurations, or device malfunction. Each controller sends periodically (every 300 s) a list of 26 data objects (e.g., power consumption, voltage, power factor, etc.) to the IoT server. A temporal series

database stores data from all controllers. This module accesses this database for the following:

- Detection of anomalies between the variables from one controller. In particular, detecting incoherency between the set of 26 read variables. This may be useful for detecting malfunction or unbalanced power lines. The AI technique used was principal component analysis trained with a sufficient number of components so that anomalies could be detected without false positives after performing Q residue control over a new input vector.
- Detection of anomalies in the time series of each controller. This is useful for detecting patterns such as unexpected power consumption (e.g., consumption during daylight). We examine the per-line power consumption for abnormal values against normal behavior (which is corrected using the daily sunset/sunrise time, added to each data sample). We have trained two models of normality (for sunset and sunrise, respectively) using 24 hours of data from the temporal series. The AI technique developed was a hybrid model that uses Extended Isolation Forest and K-nearest neighbors.
- Detection of operational issues such as power overconsumption or device malfunction. For this, we compared the detected consumption received from a controller with average consumption (and variance) over a 24-hour sliding window. If the difference is greater than a certain threshold (and variance lower than a certain threshold) over a sustained period (e.g., 5 hours) we generate a warning event.
- B3: Correlator module. This module receives the events generated by modules B1 and B2. It adds intelligence to infer the type of attack or issue under progress, figuring out the root cause and generating the final alarm messages sent to the SOC. It is based on the lightweight correlation engine SEC.

The data collected from real-world deployments includes more than 60 light controllers during three months of operation. We have used two weeks of this data set for training, another two weeks for testing, and two months for validation. The designed test includes (a) insertion of network attacks (e.g., port scanning, DoS, man-in-the-middle) and (b) contamination of samples in the data set: malicious behavior such as turning off lights during nighttime or its counterpart. We have detected several anomalous situations which have been validated with experts. Our system scored 100% of detection capacity and a rate of false positive of 10 alarms in two months. The computational resources used by our system were moderately low, enabling the monitoring of up to

600 controllers with a single conventional computer (CPU i7, 8GRAM) which exceeds the need of most deployments (usually between 60 and 200 controllers). The next step is testing in a real-work deployment.

ACKNOWLEDGMENTS

This project has been carried out in collaboration with Wellness Techgroup and with the support of the following researchers: Jesús Esteban Díaz-Verdejo, Agustín Walabonso Lara Romero, and Germán Madinabeitia Luque.

WORKS CITED

Bhamare, D., Zolanvari, M., Erbad, A., Jain, R., Khan, K., & Meskin, N. (2020). Cybersecurity for industrial control systems: A survey. *Computers & Security*, *89*, 101677.

Cheminod, M., Durante, L., & Valenzano, A. (2012). Review of security issues in industrial networks. *IEEE Transactions on Industrial Informatics*, *9*(1), 277–293.

Knowles, W., Prince, D., Hutchison, D., Disso, J. F. P., & Jones, K. (2015). A survey of cyber security management in industrial control systems. *International Journal of Critical Infrastructure Protection*, *9*, 52–80.

3.2 RELIABLE GOVERNMENT AUTOMATION OF REGULATED INFRASTRUCTURES BY SERVICE LEVEL AGREEMENTS

Pablo Fernández Montes and José María García

ABSTRACT

Public administration plays a leading role in providing adequate IT services to citizens and the need for accountability represents a fundamental principle; in particular, this need is of utter importance in the governance of IT infrastructures that support the general dynamics of administration. In this paper, we present the roadmap for a new generation of IT governance platforms that improve the level of transparency using blockchains while having a performant behavior. In addition, as a first step, we provide a high-level overview of the idea of Elastic Smart Contract as a novel element that addresses the analytical challenges present in IT governance.

INTRODUCTION

Public administration has a fundamental role in providing appropriate IT services to citizens, which would drive the increasing need for the digitalization of society. In such a context, the need for accountability represents a fundamental principle that should be intertwined into the layers of public organizations, but it is especially important in the IT infrastructures that support the general dynamics of administration.

To address a certain degree of accountability and increase transparency in IT systems, in recent years blockchain technology has become an appropriate choice to evolve current systems and incorporate nontampering mechanisms in distributed scenarios. In this paper, we present the roadmap toward a new generation of IT governance platforms that take advantages of these new possibilities to improve the level of transparency, while having an efficient behavior in the governance. In particular, we outline promising first results with the development of the concept of Elastic Smart Contract which could be used as the framework to drive analytics that supports automated governance while maintaining appropriate levels of performance, non-manipulation, and transparency.

STATE OF THE ART

Public administration represents a complex scenario where a variety of stake-holders have to collaborate, sharing information between them and allowing each party to carry out analysis and provide decentralized services on shared data. Trust issues become critical in this environment, as multiple parties (one being the citizen) have to continuously agree on the validity of the data and services they need to integrate. Blockchain technologies fit naturally, providing transparency and non-alteration to shared data in a trustless network. In addition to these features, privacy and rights management can be considered by using different blockchain implementations, ranging from permissioned blockchains (Androulaki et al., 2018) to specific solutions tailored to IoT-based ecosystems (Dorry et al., 2017), including novel approaches to data management that focus on trust and privacy preservation (Zhaofeng et al., 2020).

In recent years, the novel element concept of Smart Contract has appeared to extend the blockchains as transaction placeholders to introduce a more pro-active network and extend its capabilities; specifically, Smart Contracts represent a framework to develop a computational mechanism combining off-chain data with the one present in the blockchain.

Since the introduction of Smart Contracts (Buterin, 2013), blockchains have evolved from mere distributed digital ledgers to distributed computing platforms that can include not only an immutable data repository but also logical and behavioral information to automatically rule the relationships between stakeholders. Thus, Smart Contracts can encode functionality needed to provide additional services on top of the data registered in the blockchain. These contracts essentially aggregate some data under certain conditions that will trigger their execution. Although the data used within the contract logic is mostly obtained from the blockchain where the contract is deployed, oftentimes, there is a need to consider external data (commonly referred to as off-chain data). To preserve the untrustworthy characteristic of blockchains, an additional agent, namely an oracle, needs to provide the external data in a secured, trusted form.

CONTRIBUTION

The context of the current work has been the ANA project (Reliable Government Automation of Infrastructures Regulated by Service Level Agreements) where we have extended the pre-existing platform Governify, deployed in the Andalusian regional administration to support the IT governance in a concrete set of departments. Specifically, Governify is a service agreement management framework that boosts service governance by supporting audits in an automated way. It is composed of a set of integrated components that can be combined to create configurable architectures that adapt to each scenario. The governance platforms built with Governify gather evidence from multiple external sources in the organization (by means of their APIs) and provide visual dashboards to understand the current risks of not meeting targets. The Governify underlying agreement model (iAgree) provides a uniform modeling approach in a wide range of domains: from Service Level Agreements (SLA) in RESTful services to Service Objective/Penalties and Rewards in IT Service Support Desks driven by humans, or Best Team Practices in Agile Development Teams. These holistic capabilities allow the definition of integrated metrics, goals, and dashboards to create a common governance platform to drive the strategy of the organization. From a technological standpoint, Governify provides the native microservice architecture of RESTful components that can be easily deployed and operated as containers in the chosen infrastructure.

Nowadays, APIs are considered new business products and an increasing number of organizations are publicly exposing their APIs to create new business opportunities in this so-called API economy. In the case of the Public Administration, there is also a similar trend to adopt the so-called microservices architecture where the multiple information systems provide APIs to conform to a large ecosystem of integrated services. In such a context, defining the expected SLAs of API (including elements such as the quotas or fees for the different stakeholders) is becoming a crucial activity in the general governance of the platform. In particular, within a joint and collaborative work with the company EVERIS, we have addressed these challenges in the context of a widely used mobile application that helps citizens by aggregating in a single point, multiple services from the wide list of departments that conform to the administration. In such a scenario, that variety of services has a direct correspondence with an underlying layer of APIs that are provided by multiple scattered infrastructures that make up a wide and complex distributed scenario.

In order to improve the transparency in this scenario, we have extended the Governify platform to create a transparency layer that stores and develops analytics in a non-tampered way using blockchain technologies and the Smart Contract paradigm. Specifically, we can define two different kinds of data used as input for analysis (i.e., by means of Smart Contracts) in the blockchain paradigm. On the one hand, the paradigm provides a persistent, immutable, and non-tampered way to store a set of transactions in the chain. On the other hand, for the sake of efficiency, in actual implementations of the paradigm, there are also other current (and mutable) data available in the ledger that can be used in analytics (such as the objects in Hyperledger Fabric). In such a context, although accessing and modifying that mutable data are highly efficient, as the global size of that kind of data increases, there could be a severe impact on the performance of the blockchain. Consequently, the performance implications of maintaining a large data set impose a trade-off on the appropriate size of data to be kept for the analysis while maintaining appropriate blockchain performance. This trade-off represents an important challenge that we address by proposing a new innovative concept: the Elastic Smart Contract.

Elastic Smart Contracts represent an extension of the pre-existing paradigm to incorporate an automated orchestration of data management and analytical transactions into the blockchain to avoid saturating the blockchain and maintain the acceptable overall performance of transactions. Our preliminary results show that this new model could be adapted to variable situations to have an automatic adaption of the analytics and support the challenges presented in the IT governance of infrastructures.

ACKNOWLEDGMENTS

This project has been carried out in collaboration with EVERIS and within the ISA Research Group of Applied Software Engineering.

WORKS CITED

Androulaki, E., Barger, A., Bortnikov, V., Cachin, C., Christidis, K., De Caro, A., Enyeart, D., Ferris, C., Laventman, G., Manevich, Y., Muralidharan, S., Murthy, C., Nguyen, B., Sethi, M., Singh, G., Smith, K., Sorniotti, A., Stathakopoulou, C., Vukolić, M., Weed Cocco, S. & Yellick, J. (2018). Hyperledger fabric: A distributed operating system for permissioned blockchains. In *13th EuroSys conference* (pp. 1–15). Association for Computing Machinery.

Buterin, V. (2013). A next-generation smart contract and decentralized application platform. *Ethereum.org. Tech. Rep.* https://ethereum.org/en/whitepaper/

Dorri, A., Kanhere, S. S. & Jurdak, R. (2017). Towards an optimized blockchain for IoT. In *Second international conference on Internet-of-Things design and implementation* (pp. 173–178). Association for Computing Machinery.

Zhaofeng, M., Xiaochang, W., Kumar Jain, D., Khan, H., Hongmin, G. & Zhen, W. (2020). A blockchain-based trusted data management scheme in edge computing. *IEEE Transactions on Industrial Informatics*, *16*(3), 2013–2021.

3.3 GENERATING DATA SETS FOR INTENT CLASSIFICATION IN CONVERSATIONAL SYSTEMS

Ricardo Durán Viñuelas and Manuel Castro Malet

ABSTRACT

Generating a new conversational system requires a large data set, which is something expensive to create. Thus, many authors are adopting the use of crowdsourcing techniques for the elaboration of these data sets. However, this process requires a significant amount of manual effort related to the annotation and validation of the input data. We adopt the use of Artificial Intelligence and Machine Learning techniques that considerably reduce the manual workload and ensure a quality data set.

INTRODUCTION

In recent years, task-oriented dialog systems have been growing steadily in home and work environments. Moreover, platforms and tools such as Google's DialogFlow or Rasa have been made available to any user, company, or industry, helping the fast development of new chatbots. However, dialog systems require large high-quality data sets to ensure accuracy, which are extremely expensive to build, measured in both annotator-hours and financial cost.

Some of these issues can be solved using crowdsourcing tools like Amazon Mechanical Turk (AMT), a significantly cheaper and faster method of collecting annotations or data sets, thanks to a large number of non-expert contributors from anywhere in the world, doing the work in exchange for micropayments via the web. Besides incorporating a large number of workers as temporary labor, crowdsourcing techniques are quite appropriate for the task of generating data sets, because they also provide a diversity hardly achievable by a reduced group of workers. On the other hand, it forces a task of supervision of the results since a significant number of tasks may contain errors.

In addition to crowdsourcing techniques, the automation of tasks such as classification, labeling, and data processing can also reduce the amount of human effort. In this sense, the recent proposals applying transformer-based architectures over Machine Learning frameworks such as TensorFlow have proved to increase performance significantly.

In this work, we propose a methodology to build a data set using AMT and Machine Learning techniques that minimizes human efforts and ensures a quality data set.

STATE OF THE ART

The quest to reduce the large human effort involved in creating a complete data set, in addition to the cost, is a recurring topic in research. Many authors have studied the validity or benefits of using AMT. In April 2021, the Scopus database included more than 2,500 stored articles that talk about the use of AMT.

The work of Sabou et al. (2014) is the starting point for proposing a crowdsourcing approach for the creation of a data set, as it summarizes and proposes guidelines and good practices in crowdsourcing projects. The crowdsourcing process in the field of collecting large amounts of data can be broken down into four main stages: *data definition*, *data preparation*, *project execution*, and *data evaluation and aggregation*. Traditionally, many of these stages have been performed manually.

Kang et al. (2018) present two interesting metrics, *diversity* and *coverage*, and evaluate the performance of the scenario-driven and paraphrase methods and their variants when collecting data for a production dialogue system. Following this article, Larson *et al.* (2019) build a data set for intent classification, needing a large human effort to classify and evaluate the data obtained through AMT. As shown in the "Contribution" section, we take these two articles as the main references and propose a new approach that tries to automate these tasks reducing the human effort carried out by the authors.

On the other hand, there are models that can be used to compute sentence-level semantic similarity scores leading to the automation of text classification, clustering, or other natural language tasks (Cer et al., 2018). Semantic similarity is a widely useful measure in several natural language processing tasks.

Along the same lines of automating the laborious manual classification work, a recent work has proposed the use of DBSCAN, a popular density-based clustering algorithm that searches for clusters (Chatterjee & Sengupta, 2021). The results obtained by the authors are encouraging.

CONTRIBUTION

When facing the development of a new conversational application, two scenarios can be considered: (1) Scenario 1: one or more sub-domains or topics are provided or (2) Scenario 2: a specific set of intents is requested, with a description and some examples provided for each intent.

Agreeing with the approach proposed by Larson *et al.* (2019), we split the work needed to solve both scenarios into two different tasks: (1) Task type_1: converting a topic or domain into a set of intents with example sentences for each intent and (2) Task type_2: expand the coverage of an individual intent using *rephrase* and *scenario* tasks proposed by Kang et al. (2018).

The resolution of Scenario 1, in which the intents covering the application domain are unknown a priori, would be achieved by executing M tasks type_1 followed by $\sum_{i=1}^{M} N_i$ tasks type_2 (where M is the number of topics and N_i is the number of intents identified for the topic M_i). For Scenario 2, the resolution would be performed by P tasks type_2 (where P is the number of intents specified).

Several articles emphasize a large amount of manual workload that is usually associated with the resolution of this type of tasks despite the use of crowdsourcing. This workload is manifested in three main points: (a) generation of Human Intelligence Task, that is, task definition and data preparation, (b) validation of task results, taking into account the potentially high number of tasks with errors due to the use of crowdsourcing, and (c) the subsequent work of annotation or classification of the results (applicable to Scenario 1).

Therefore, our approach proposes the following methodology:

Task type_1: given a domain or topics, get sample sentences for each intent as follows (1) manual work consisting of the definition of the task and the domain or topics and launching the task in AMT, (2) we accept all tasks from AMT workers, even though we know that some tasks may not have been performed correctly. To reduce the number of errors, we use data preprocessing cleaning techniques such as spellchecking or language detection models, (3) in order to group the sentences received in intents, we propose to use clustering techniques, such as those indicated in Chatterjee and Sengupta (2021) using the distances calculated by Universal Sentence Encoder (Cer et al., 2018). In this way, we obtain several clusters of sentences separated from each other, where each cluster corresponds to a different intent, and (4) we validate the data after the clustering stage. Although it is not automatic, its impact is minimized using the two following points. We will look for (a) clusters with less than n example sentences, which is indicative of sentences that may not belong to the domain at hand and can be potentially deleted or (b) within a cluster or intent, the LOF (Local Outlier Factor) is checked to identify and discard outliers.

Task type_2: expand the data set for each of the intents given or identified as follows (1) following the work of Larson et al. (2019) and Kang et al. (2018), we use the *rephrase* and *scenario* tasks to collect additional data for each intent again in AMT. The definition of the *scenario* task requires a manual process, unlike the *rephrase* task which is automatic, since the sentences to be paraphrased are the sentences themselves already obtained from the previous task, (2) once the results from AMT are received, we repeat the data preprocessing cleaning process using the same techniques and methods previously used in task type_1, and (3) we validated the data by re-performing the clustering techniques and rechecking the LOFs. For every sentence received, the semantic distance for every intent is calculated, and we identify if the sentence should be included in the proposed intent, or it should be discarded.

ACKNOWLEDGMENTS

This project is carried out in collaboration with 4i Intelligent Insights SL and within the Julietta research group in natural language processing.

WORKS CITED

Cer, D., Yang, Y., Kong, S., Hua, N., Limtiaco, N., St John, R., Constant, N., Guajardo-Céspedes, M., Yuan, S., Tar, C., Sung, Y.-H., Strope, B., & Kurzweil, R. (2018). Universal sentence encoder. *EMNLP demonstration*. Association for Computational Linguistics, Brussels, Belgium.

Chatterjee, A., & Sengupta, S. (2021). Intent mining from past conversations for conversational agent. In *Proceedings of the 28th international conference on computational linguistics, 2020*.

Kang, Y., Zhang, Y., Kummerfeld, J. K., Hill, P., Hauswald, J., Laurenzano, M. A., & Tang, L. (2018). Data collection for a production dialogue system: A clinic perspective. In *16th annual conference of the North American chapter of the association of the computational linguistics (NAACL)*, New Orleans, LA.

Larson, S., Mahendran, A., Peper, J. J., Clarke, C., Lee, A., Hill, P., Kummerfeld, J. K., Leach, K., Laurenzano, M. A., Tang, L., & Mars, J. (2019). An evaluation dataset for intent classification and out-of-scope prediction. In *Proceedings of the 2019 conference on empirical methods in natural language processing and the 9th international joint conference on natural language processing* (pp. 1311–1316), Hong Kong, China.

Sabou, M., Bontcheva, K., Derczynski, L., & Scharl, A. (2014). Corpus annotation through crowdsourcing: Towards best practice guidelines. In *Proceedings of the 9th international conference on language resources and evaluation*, 859–866.

3.4 AN ARTIFICIAL INTELLIGENCE TOOL TO CALIBRATE THE COVERAGE OF INFORMATION SOURCES ON THE INTERNET

Nuria Gómez-Vargas and Jasone Ramírez-Ayerbe

ABSTRACT

This project focuses on the development of tools for the automatic updating of the *Directorio de Empresas y Establecimientos en Andalucía*. Artificial Intelligence techniques are used to analyze the different information sources found on the internet. Concretely, a process is developed to automatically retrieve the content of the different websites and calibrate their data quality, for its later use to update the features of the directory in the case of high scores. Furthermore, our procedure makes it possible to detect new registrations and deregistrations of establishment, as well as to evaluate the representativeness of the different sectors on the internet.

INTRODUCTION

The updating of the *Directorio de Empresas y Establecimientos en Andalucía* involves contracting private sources and accessing administrative registers every year, hence a study of alternative sources and the development of automatic information retrieval procedures are highly needed. The works carried out within the framework of this project have been focused on the identification of complementary information sources, the development of tools for automatic information retrieval from selected sources, and the analysis of their data quality, to contrast them with the information available in the directory.

The data analysis and comparison are based on the linkage with the existing information, using either the normalized literal information or the geographic location, depending on whether companies or establishments are being considered, respectively. With the information from those sources valued as reliable, the dynamic update of the variables that make up the different registers is accomplished. In addition, this process allows the detection of registrations and deregistrations of establishments, as well as assessing the representativeness of the different sectors on the internet.

For this, a series of procedures based on Artificial Intelligence techniques have been defined, which has allowed the automatic assessment of the sources for their subsequent exploitation.

STATE OF THE ART

Web scraping is an important technique used to extract unstructured data from websites and transform it into structured data that can be stored and further analyzed in a database. An overview of this technique and its tools can be found in Saurkar et al., 2018. These routines are used in diverse areas like product price comparison, weather broadcasting, advertisement and market analysis, or the extraction of business details from business directory websites. In the latter field, we can find some research (Rhodes et al., 2015, Kim et al., 2016) in which the aim is not only to obtain information about establishment using web scraping but also to compare and contrast the information obtained from different sources.

The correct verification of the retrieved data is of great importance and because of this, in the aforementioned research, the veracity of the scraped information is checked through a crowdsourcing platform, which had workers contacting the establishment to verify each record. However, this routine task is costly, time-consuming, and also subject to possible human error. For this, we propose as an innovative alternative a calibration method, using Artificial Intelligence techniques, that does not require this manual work.

Although there are references in the literature of several studies using the online search methodology on establishments and companies, there is no previous research that we are aware of that addresses the dynamic comparison with a verified directory for the updating of its diverse registers. Only evaluations of data quality standards have been found (Cai et al., 2015) but they focus on business needs in the sense of analyzing indicators (auditability, relevance, accessibility, etc.), and the processing of big data. This is why we consider our work to be innovative and a major contribution, due to the computer-based strategy that addresses the study of the data quality based on both the normalization of addresses and the standardization of the respective geolocation of these.

CONTRIBUTION

The starting point of this study is the different information resources found on the internet that meet the requirements defined by the Instituto de Estadística y Cartografía de Andalucía (IECA). These websites have been crawled to define narrow searches that automatically filter out only the requested data from the establishments or companies and retrieve them using web scraping techniques. This information might be given in the form of geographic coordinates or character strings. In the first case, for the comparison with the directory data, the coordinates have been standardized; meanwhile, for the second data type, it has been necessary to use aLink, the normalization tool provided by IECA.

Once the information has been scraped and tuned-up, different methodologies have been established depending on the nature of the information—according to if they belong to companies or establishments—in order to carry out the dynamic updating of the variables that make up the records of the "Directorio de Empresas y Establecimientos en Andalucía".

In the case of companies, a supervised classification has been defined since each instance has a unique identifier, the TIN (Taxpayer Identification Number). This code has been used, in a first step, to make a direct link between the directory and the database with the companies found on the internet. In a second step, the algorithm addresses the non-matches based on the information that has been taken as truthful, the one in the directory. For instance, the number of non-coincidences decreases when inactive establishments are checked based on the updated information date, and a wider evaluation can be assessed due to the massive scraping, by inspecting companies that have been retrieved but with a National Economic Activities Classification Code associated with another code within the same sector. Once the coincidences have been matched, the quality of the source is analyzed by comparing the addresses associated with the same company, assigning a distance value between 0 and 1 to each match. First, to optimize comparisons, another unique identifier is used and the worst case (a distance equal to 1) is assigned to addresses belonging to different postal codes. For the ones which match, the string distance between the normalized street names is computed. Address normalization is a complicated process, thus obtaining the same literals in both records (a distance equal to 0) is rare. Because of this, the literal information has been preprocessed by deleting from each record the so-called "stop words". This term refers to the most common words in a language, like articles or prepositions, whose position within the literals depends on the normalization performed. The distance between strings might highly depend on the order of the words, therefore by ignoring them and retaining only the most significant parts of the literal, more accurate distances are obtained. There are different distances to compare string literals, such as Levenshtein or Jaro-Winkler. Finally, data quality evaluation is assessed by analyzing the histogram of the chosen distance, as well as the proportion of extreme values 0 and 1.

Regarding the information of establishments, the exact geographic position has been obtained. In this case, a process has been developed including a coordinate conversion step and the implementation of the haversine formula expressed in terms of a two-argument inverse tangent function to calculate the distance between two points. This procedure enables the linkage of the scraped addresses to their nearest one from the directory. However, the latter are geographically localized using the "Callejero Digital de Andalucía Unificado" (CDAU), which does not always have the information of the exact doorway and thus links the address to the center of the road or other approximations. That is why a more complex process is used for the allocation of scores: different considerations are taken into account to correctly establish the matches, such as the type of geolocation in CDAU (exact portal, nearby or via center), the previously introduced distance between the linked addresses according to a threshold and the number of establishments per street. Finally, a score is associated with each instance using all the above features.

The results of this analysis allow, in the case of instances with sufficiently high scores, the dynamic and automatic update of the directory based on information scraped from the internet.

ACKNOWLEDGMENTS

This project has been carried out in collaboration with the Institute of Statistics and Cartography of Andalusia and with the support of the following researchers: Rafael Blanquero (IMUS), Elisa Isabel Caballero (IECA), Emilio Carrizosa (IMUS), Marina Enguidanos (IECA), and Gema Galera (IECA).

WORKS CITED

Cai, L., & Zhu, Y. (2015). The challenges of data quality and data quality assessment in the big data era. *Data Science Journal, 14.* DOI: http://doi.org/10.5334/dsj-2015-002

Kim, A. E., Loomis, B., Rhodes, B., Eggers, M. E., Liedtke, C., & Porter, L. (2016). Identifying e-cigarette vape stores: Description of an online search methodology. *Tobacco Control, 25(e1),* e19–e23.

Rhodes, B. B., Kim, A. F., & Loomis, B. R. (2015). Vaping the Web: Crowdsourcing and web scraping for establishment survey farm generation. In *Proceedings of the 2015 federal committee on statistical methodology research conference.* December 1–3, 2015, Washington DC.

Saurkar, A. V., Pathare, K. G., & Gode, S. A. (2018). An overview on web scraping techniques and tools. *International Journal on Future Revolution in Computer Science & Communication Engineering, 4*(4), 363–367.

Mobility Logistics and Advanced Industry Linked to Transportation

4

For a region like Andalusia with an area of more than 87,000 km,[2] improvements in the efficiency of transport, logistics, and mobility are key to its productivity. Research groups are aware of these needs and have very active collaborations with these sectors. Examples can be seen in improvements in the management of goods in warehouses with collaborative inventories in retail, improvements in the maintenance of urban passenger buses, rail transport, and pipeline networks, allowing the digital transformation of factories of small and medium size, the applications to the aerospace sector, or the use of image sensors that improve intelligent movements.

This chapter presents an example of the use of optimization of Machine Learning techniques to improve the replenishment of fashion stocks for retailers. Textile companies are evolving with increasingly shorter seasons, having to continually change collections to offer customers new value-added products. The proposed logistics management system is valid for any distribution and supply chain with nonperishable stocks that lose their value over time and is particularized for the main warehouse and several stores. Its purpose is to dynamically optimize the inventory of the stores serving a wide range of products, based on a sales forecast from previous periods.

Likewise, the development of a surveillance system for the maintenance and diagnosis of urban buses is presented, using a model of the urban-bus-cooling system based on neural networks, for the diagnosis of the system, and for the monitoring of bus engines which run many miles a year based on data transmitted wirelessly. It will improve the efficiency of the predictive maintenance of different elements of the buses, avoiding breakdowns in operation and improving their operability.

DOI: 10.1201/9781003276609-4

Another real problem in which the current situation is being investigated and improved is in the routing of ship pipelines using mathematical optimization and a Machine Learning approach. Currently, this problem has a difficult solution as the space of the ships through which the route of the pipes has to pass is limited, opting to solve it with Artificial Intelligence heuristics.

The same importance has the digital transformation that is coming in factories, proposing Industry 4.0 models to transform them into digital factories that are accessible with investments for small- and medium-sized companies.

Furthermore, a paper applied to the aerospace sector is presented using Machine Learning to ensure the harshness of the radiation suffered by electronic systems in space, ensuring their performance and functionality during their useful life. Radiation degradation in components used for commercial devices is studied, evaluating the risk associated with these radiation environments.

In addition, the importance of rail transport in the movement of people and goods is presented, with an automatic learning system that feeds on the information captured by techniques that will facilitate the conservation of this type of railway infrastructure. This learning system will allow the early detection of failures, which help predict the evolution of the state of use, estimate maintenance operations, and plan them.

Finally, an application of Artificial Intelligence techniques is presented that improve analog image sensors in energy consumption, having an important use in the efficiency of semiconductors and nanoelectronics. With these improvements, the robots will be allowed to detect small changes in the environment, when interacting with arbitrary stimuli.

4.1 OPTIMIZING STOCK REPLENISHMENT IN FASHION RETAILERS

Jesús Muñuzuri Sanz and Alicia Robles-Velasco

ABSTRACT

In the current paradigm, where rapid evolution of the retail industry is seen, especially in textile companies, general product assignment in any distribution and supply chain, consisting of a main warehouse and several locations, can mean an important challenge. The replenishing process in shops centers all its attention on dynamically optimizing the inventory of shops attending to a wide range of products, starting from a sales prevision of previous periods. In this context, the objective of this work is centered on optimizing replenishing work in shops for non-perishable goods which lose value as time passes.

INTRODUCTION

Replenishing in shops must satisfy independent flow demands, in methods and time, allowing covering the demands of shops without increasing their associated warehousing costs. Different sectors (for example, textile) have specific casuistry facing a shortage of any product. In normal conditions, this is penalized, but on certain occasions, this effect is not that subject to accusation because customers opt for different equivalent products from another shop of the same company or they acquire them via web channels. Nevertheless, the fashion sector is subject to the strong depreciation of the product (end of the season), because it has not been sold during the season for which it was designed. This means that unsold stocks are available at prices with large discounts, greatly reducing the profit margin.

The objective of this work is to optimize the replenishment in shops, guaranteeing maximum product availability and rotation, and so, reducing costs, especially those derived from excess storage or, on the contrary, the costs of stock shortages. Therefore, a version conjugating two different metaheuristics has been implemented: particle swarm optimization (PSO) for updating solutions and Simulated Annealing to calculate the changes experienced in the solution, in addition to limiting simulation time.

STATE OF THE ART

Currently, the shop replenishment problem has been repeatedly discussed to improve the planning and optimization of the supply chain and retail commerce. Nevertheless, only some researchers, such as Iannone et al. (2013), focus their attention on the textile sector and all its casuistry: extensive product and customer catalog, high rate of rotation, highly unpredictable, as well as seasonal and impulsive demand.

Pan et al. (2009) consider batch economic models subject to market changes or seasonality. Along this line, Grewal, Enns, and Rogers (2015) assess the sensitive price variation dominated by the randomness and seasonality of demand. Searching for demand prediction, Anily and Hassin (2013) focus on heterogeneous behavior of customer decision. Outside of the textile sector, other works, such as Coelho and Laporte (2014) who focus on the reestablishment and replenishing of perishable or deteriorating products in shops should be highlighted.

Lee (1987) calculates the minimum cost of an inventory with transfers for replacement n shops, located in the same territorial area, optimizing the expected number of pending orders. Later, Kukreja et al. (2001) add the casuistry of shops located in different territorial regions. Rudi, Kapur, and Pyke (2001) study a periodic review system with the restocking of two shops after knowing the demand for the period but before it is satisfied. On the other hand, Archivald (2007) develops an inventory system based on a multi-location model with periodic reviews for different industrial environments. He focuses on quantifying the optimal decisions of transfers between locations according to the moment in which movements are allowed.

On the other hand, Minner, Silver, and Robb (2003) develop heuristics to define the convenience or not of restocking or moving items in a continuous multi-location review inventory where shortages are allowed. Elaborating along this line, Axsäter (2003) implements a replenishing policy where the shop with a shortage can obtain products from other shops that have less stock shortage costs. The paper of Martino, Iannone, and Packianather (2016) has been used for reference in this development, as a guide to define the model and its equations, as well as the study of the shop replacement problems.

CONTRIBUTION

The issue contemplated in this study is about shop inventory and replenishment management for a multi-product model with multiple locations and over various time periods. The main objective of the problem is to maximize profitability, understood as the difference between total income and costs. Among the main considered hypotheses, we can mention the following:

- Each shop will receive an order from the central warehouse at the start of each period. The order will consist of different typologies of products.
- The periods can be defined as a range of days, weeks, months, or years.
- This means that the products are not perishable but do suffer from devaluation over time.
- There is an estimated demand prevision based on historical data.

Revenue has been defined as the unit sale price of each product in each shop during each period. The sales of each product and in each shop will be estimated as the minimum value between the inventory existing in the previous year and the demand in the current period. The demand of each product, in each shop and for each period, will be a value between the range of sales predictions with an uncertainty threshold of up to 20%.

In terms of costs, stock shortage costs can be considered only if the demand to be satisfied is greater than the inventory existing in the previous period; otherwise, it will be nil. The purchase cost is the cost of acquiring all the units of the different orders for each shop and period. The transport cost is considered as separated into two concepts: the fixed-route cost of the vehicle to supply all the shops and the variable cost depending on the distance and type of order to be covered. This means that each vehicle travels back to one shop without visiting any others. Finally, the storage cost is the sum of a fixed inventory maintenance cost per shop and a variable cost evaluable by the depreciation of the inventory during the period.

Therefore, the proposed optimization problem consists of maximizing the benefit understood as the difference between income and costs, on the condition that purchase costs do not exceed a previously set budget. Artificial Intelligence techniques were used to solve the problem. Specifically, the developed methodology combines PSO metaheuristics where a leader (best solution) sets the direction of solutions group, with Simulated Annealing for upward movements and escaping from local optima, accepting or denying the changes produced in the solution.

This typology of problems presents many feasible solutions. However, with a finite period, there is no guaranteed convergence with the optimum. Therefore, the best solution until meeting a fixed stop criterion is sought in a maximum number of iterations.

The sequence of steps of this methodology is as follows: initially, a demand, a random period, and a current solution are calculated, defining the sales and the inventory for that first solution, which will be the origin of the first vicinity. All the neighbors of vicinity are formed based on a percentage variation of values and periods. This variation can be incremental or decremental in the range between 0% and 40%.

Subsequently, each solution will be valued defining its benefit, and the best solution will be updated according to its value. If the change of the current benefit is less than a limit, that change will be accepted updating the current solution. The creation of a new series of solutions has been designed as the weighted measure of the best solution found so far, the current solution and a random solution (the used weights have been fixed by experimentation).

ACKNOWLEDGMENTS

This project has been carried out in collaboration with the TIER1 Technology and within the Organization Engineering research group.

WORKS CITED

Anily, S., & Hassin, R. (2013). Pricing, replenishment, and timing of selling in a market with heterogeneous customers. *International Journal of Production Economics*, *145*(2), 672–682.

Archibald, T. W. (2007). Modelling replenishment and transshipment decisions in periodic review multilocation inventory systems. *Journal of the Operational Research Society*, *58*(7), 948–956.

Axsäter, S. (2003). Evaluation of unidirectional lateral transshipments and substitutions in inventory systems. *European Journal of Operational Research*, *149*(2), 438–447.

Coelho, L. C., & Laporte, G. (2014). Optimal joint replenishment, delivery and inventory management policies for perishable products. *Computers & Operations Research*, *47*, 42–52.

Grewal, C. S., Enns, S. T., & Rogers, P. (2015). Dynamic reorder point replenishment strategies for a capacitated supply chain with seasonal demand. *Computers & Industrial Engineering, 80,* 97–110.

Iannone, R., Ingenito, A., Martino, G., Miranda, S., Pepe, C., & Riemma, S. (2013). Merchandise and replenishment planning optimisation for fashion retail. *International Journal of Engineering Business Management, 5,* 5–26.

Kukreja, A., Schmidt, C. P., & Miller, D. M. (2001). Stocking decisions for low-usage items in a multilocation inventory system. *Management Science, 47*(10), 1371–1383.

Lee, H. L. (1987). A multi-echelon inventory model for repairable items with emergency lateral transshipments. *Management science, 33*(10), 1302–1316.

Martino, G., Yuce, B., Iannone, R., & Packianather, M. S. (2016). Optimisation of the replenishment problem in the Fashion Retail Industry using Tabu-Bees algorithm. *IFAC-PapersOnLine, 49*(12), 1685–1690.

Minner, S., Silver, E. A., & Robb, D. J. (2003). An improved heuristic for deciding on emergency transshipments. *European Journal of Operational Research, 148*(2), 384–400.

Pan, A., Leung, S. Y. S., Moon, K. L., & Yeung, K. W. (2009). Optimal reorder decision-making in the agent-based apparel supply chain. *Expert Systems with Applications, 36*(4), 8571–8581.

Rudi, N., Kapur, S., & Pyke, D. F. (2001). A two-location inventory model with transshipment and local decision making. *Management Science, 47*(12), 1668–1680.

4.2 DEVELOPMENT OF A SURVEILLANCE SYSTEM FOR MAINTENANCE AND DIAGNOSIS OF BUSES BASED ON CAN-BUS DATA TRANSMITTED WIRELESSLY

Francisco José Jiménez-Espadafor Aguilar and
Juan Manuel Vozmediano Torres

ABSTRACT

From the point of view of vehicle maintenance, one of the most important systems of urban buses is the cooling system. These vehicles run typically more than 80,000 km per year, and the radiator of the system gets fouled due to dust and dirt of the cooling air, which produces an increase in water temperature. This situation forces to stop the vehicle and perform washing of the radiator. This study is focused on the development of a model of the cooling system of urban buses based on an artificial neural network (ANN), which is used for system diagnosis and engine surveillance. Data are gathered from the CAN-bus system of every bus, which have allowed the development of a dynamic ANN that fits the cooling dynamics.

INTRODUCTION

For urban buses, energy consumption comes from the propulsion system, the auxiliaries, and the thermal management. The latter include the electric devices, the generator, and the cooling of the internal combustion engine (ICE). In this regard, maintenance of radiator takes out from service the bus and requires many hours of work and, therefore, is expensive from the maintenance point of view. If this issue is considered in a medium city transport company as the one of TUSSAM in Seville (Spain), with a fleet of more than 400 urban buses, the problem highlights because currently there is no procedure able to determine the state of the radiator.

STATE OF THE ART

From the point of view of energy consumption, thermal management is very relevant next to the rest of the energy fluxes inside a propulsion plant. The importance of considering thermal management as a procedure to reduce fuel consumption in ICE has been demonstrated (Caresana, Bilancia & Bartolini, 2011), where performance is not only to maximize fuel efficiency but to increase system life and reduce the maintenance cost (Bayraktar, 2012).

CONTRIBUTION

Cooling System

ICE has to operate below a temperature limit in order to guarantee performance and reliability. For this reason, the cooling system has to be properly maintained for the dissipation of the thermal energy generated by the whole vehicle. On the other side, the cooling system requires electrical energy for its duty. Therefore, the correct maintenance of this system contributes to the reduction of the energy consumption of the bus.

The cooling system includes two pumps arranged in series, a turbocharger heat exchanger, water radiators, a hydraulic fan, and several valves. Thermal energy from ICE (compressor, engine, gearbox, and retarder) is conducted by water through radiators, where it is transferred to ambient air by the fan, which has a maximum efficiency of roughly 50%. The regulation valve is managed by the temperature measured by three thermocouples that set the adequate water flow rate for any load, in order to guarantee that the maximum outlet temperature is below the limit for any operating condition.

ANN Modeling

ANNs are widely found in fields related to the diagnosis of internal combustion engines and system modeling, concretely for the detection and quantification of failures. The main reason is that they can learn the behavior of complex systems only from observations. On the other hand, ANN can be used as an inverse model (Desbazeille et al., 2010).

In this paper, only one ANN is necessary to determine system fault identification and its level and engine state, taking into account that the configured network is allowed to reproduce the output temperature of water of the engine block, gearbox, and retarder. With this porpoise, an ANN is configured, as well as trained and evaluated, in order to obtain a mapping between input variables and the corresponding output variables (targets). In this case, the input variables are as follows: instantaneous engine speed, ambient temperature, instantaneous engine torque, and instantaneous fuel consumption. On the other side, the output variables are the instantaneous temperature of the leaving water from the engine block, the output temperature of the retarder, and the gearbox.

ANN Architecture and Training Method

Since the relationship between inputs and targets is highly nonlinear, a multi-layer network has been chosen. The transfer function in the hidden layers is the tangent sigmoid function and the function for the output layer is linear. Such a network structure is considered a universal function approximator (Hornik, Stinchcombe & White, 1989). In this case, the forward function is obtained directly from data, so the configured ANN has to learn the inverse relationship. As the number of hidden layers is concerned, one hidden layer has been chosen. The number of neurons in this layer is 19 and has been obtained by trial and error.

Regarding the training method, the Levenberg–Marquardt algorithm has been used. In this kind of training method, early stopping must be taken into account to avoid the unwanted overfitting effect. Because of this, the total sample data has been divided into three groups: training set (60% of the data), test set (20% of the data), and the mentioned validation set (20% of the data).

Input and Target Data

Through the cooling system data gathered of the CAN bus, a set of ANN inputs and targets have been specified. Such a set must be a representative sample of the engine's whole load, which includes winter, spring, autumn, and summer data. This is important due to the huge change of ambient temperature and the engine load that change the thermal load on the cooling system. Holidays have also been included because the load of the bus is greatly reduced on these days. The performance of the ANN has been evaluated by means of a set of data that is not used throughout the training process, that is, the aforementioned validation set. A useful tool for the ANN validation test is the relationship between the targets and outputs. In this case, a linear relationship has been observed, which means a regression coefficient close to 1. This means that no overfitting takes place and, as a result, it can be considered that the trained ANN has achieved favorable results.

Noise Effect

In order to verify the robustness of the dynamic ANN against noise, random noise has been added to the signals coming from the input data. Two different noise signals have been considered:

Noise characterized by mean = 0 and standard deviation = 1.5%

Noise characterized by mean = 0 and standard deviation = 3.0%

The trained ANN outputs were compared to the targets and a lack of robustness could be observed. However, after a sensitivity analysis, it has been seen that the most affected signal was room temperature, having the rest low sensitivity to noise. In this regard, both sensor accuracy and position are considered to be relevant for the robust results of the cooling model.

CONCLUSIONS

A complete procedure for modeling the cooling system of an urban bus has been developed. The method is based on the integration of instantaneous CAN-bus data and a multilayer dynamic ANN. The method presents high robustness and also showed its capability for diagnosing the health of the engine oriented to preventive maintenance and failure diagnosis.

The method will be applied to a fleet of 20 urban buses that belong to TUSSAM, where it will be used for monitoring the condition of the cooling system. A huge reduction of the fleet out-of-service period is expected due to maintenance duties and an increase in fleet availability.

ACKNOWLEDGMENTS

This project has been carried out in collaboration with TRANSPORTES URBANOS DE SEVILLA (TUSSAM) and with the support of the following researchers: Antonio Cabrera García-Doncel and Daniel Palomo Guerrero.

WORKS CITED

Bayraktar, I. (2012). Computational simulation methods for vehicle thermal management. *Applied Thermal Engineering, 36*, 325–329.

Caresana, F., Bilancia, M., & Bartolini, C. M. (2011). Numerical method for assessing the potential of smart engine thermal management: Application to a medium-upper segment passenger car. *Applied Thermal Engineering, 31*, 3559–3568.

Desbazeille, M., Randall, R. B., Guillet, F., El Badaoui, M., & Hoisnard, C. (2010). Model-based diagnosis of large diesel engines based on angular speed variations of the crankshaft. *Mechanical Systems and Signal Processing, 24*(5), 1529–1541.

Hornik, K. M., Stinchcombe, M., & White, H. (1989). Multilayer feedforward networks are universal approximators. *Neural Networks, 2*(5), 359–366.

4.3 A MATHEMATICAL OPTIMIZATION AND MACHINE LEARNING APPROACH FOR PIPELINE ROUTING

Yolanda Hinojosa Bergillos and Diego Ponce López

ABSTRACT

In shipbuilding, pipeline routing is a difficult problem as space is rather limited. This constraint and others related to obstacles, costs, legislation, or operability are considered to set the pipeline layout by means of a mathematical model. The problem is solved in an exact or heuristic way when the complexity increases when dealing with real instances, using Artificial Intelligence tools. In this work, we outline some issues and solution proposals which have been applied to a case study that shows the usefulness of mathematical tools in design engineering.

INTRODUCTION

Shipbuilding consists of a conceptual design phase, when engineers determine all the necessary construction requirements, and a basic design phase, when the equipment are adequately selected and connected by pipes or electrical circuits. Due to the limited space, more pronounced in the shipping industry than in other industrial plants, the later phase is crucial. Some constraints must be satisfied by the optimal routes such as avoiding obstacles or passing through initial and end points. Others are more flexible, such as preference zones, elbows, and material and maintenance costs and can be determined with the appropriate mathematical tools.

GHENOVA is an engineering company which delivers consulting in marine services and requires design tools based on Artificial Intelligence and Machine Learning. These tools must incorporate particular features or restrictions for ships which are not currently considered by commercial software. Among them is the so-called scheme problem. The scheme encompasses a set of design specifications regarding the number of branches that must be connected to a trunk line, valves on the branches, and some other related aspects to be considered. Once the initial design has been determined, given by the scheme conditions, the pipeline routing problem will be solved following the technical specifications.

STATE OF THE ART

The first approach to the scheme problem proposed by GHENOVA company has been addressed by mixed-integer programming (Cuervas et al., 2017). In this paper, we will focus on pipeline routing, including main lines and branches.

The pipeline routing problem was previously solved by means of three-dimensional routing algorithms. We refer the reader to Park and Storch (2002) and the references therein to see some particularities of the pipeline design in a shipbuilding domain. In that paper, the proposed algorithm combines a cell-generation method to consider the geometric aspects and the correct assignment of costs, in order to take into account the non-geometric ones. Subsequently, the path for each pipeline is chosen from a tree of combinations.

Pipeline routing has been studied in other spheres beyond ships. For instance, the problem has been addressed to design chemical plants (Guirardello & Swaney, 2005). Here, the developed procedure is based on the construction of a partial graph where node positions determine where pipes can be routed, provided that the safety restrictions and minimum distances from components are satisfied.

Traditionally, pipe routing has been studied as a shortest path problem solved using A* or Dijkstra algorithms, among others. When the problem is designed in three dimensions, it becomes more complicated. To face the problem, Min, Ruy, and Park (2020) have recently designed an ad hoc Jump Point Search algorithm. This method is performed in a three-dimensional space and provides pipe paths which are parallel to the axes.

Building information modeling (BIM) software is used during the design pipeline process. However, still there is room for improvement for those automatic tools and experienced people have to intervene in choosing the final pipe system layout (Singh & Cheng, 2021). To ease the task, many heuristics and optimization methods have been studied and developed in the field in recent years.

CONTRIBUTION

There are many aspects to consider as part of solving the pipeline routing problem by means of Artificial Intelligence techniques. These issues are, among others, the adequate definition of the solution space, the correct pipe layout satisfying a given scheme design, or the definition of an appropriate function to compare the quality of different routing solutions. Our first contributions to the field were reported in Cuervas et al. (2017). In this research, several of these technical aspects were solved following the instructions of GHENOVA, a company with which we are collaborating since 2015.

Some of the contributions arising from previous and current transfer collaborations with that company are (1) the development of a procedure that discretizes the solution space by means of nonuniform meshes considering the present physical barriers, (2) an approximation of the scheme design problem by means of a mixed-integer programming model capturing the main theoretical characteristics of the problem, and (3) a cost function to take into account the singularities in naval design, such as length, number of elbows, and zone priorities. They have been included in the objective function as both bonuses and penalties.

From this starting point, we have improved the graph-based discretization solution space. Along with the nodes contained in the primitive mesh, the new graph structure adds virtual nodes to incorporate changes of directions (elbows) in a number considerably smaller than the original one and, therefore, the size of the problems has decreased. Given the graph that defines the solution space, the adequate objective function, and the known initial and end points of the pipelines which were defined by the scheme specifications, the pipeline routing problem has to be solved. Our contributions concerning this stage focus on modeling the pipeline routing problem in an efficient way, so that the problem can be solved exactly. On the other hand, for large-sized instances, we have tested Machine Learning heuristics which provide feasible good-quality solutions sought by naval engineering.

In this respect, we have developed three Matheuristics that share a common idea: the decomposition of the general problem in subproblems which can be independently solved by flow algorithms or by shortest path algorithms. It should be noted that depending on the pipe's dimensions and shapes, the pipeline routings must keep a security distance between them and, therefore, they "compete" for common space. Thus, the decomposition of the general problem on subproblems might lead to incompatibilities that can be solved by an iterated scheme. The different Matheuristics differ in the choice of the set of subproblems, in how to solve the incompatibilities, and in how to carry out the iterated scheme.

Another important issue is the minimum required distance between consecutive elbows belonging to the same route. Accordingly, we propose a recursive algorithm to obtain the shortest paths that meet the threshold between consecutive elbows that extends the classical shortest path algorithms. Nevertheless, the complexity of this new algorithm is greater than the classical ones, so we have also defined a heuristic algorithm to be integrated into the above described Matheuristics.

At the same time, we have developed a model to exactly solve the pipeline routing problem. The problem is seen as a multicommodity flow problem which extends classical models by means of adding constraints to ensure that the pipeline routes of different services keep a security distance between them. Additionally, constraints to keep the above-mentioned distance between elbows are defined. At this moment, we are working on algorithms which are able to reduce the solution space, in order to solve the problem in this reduced solution space using branch-and-cut methods.

ACKNOWLEDGMENTS

This project has been carried out in collaboration with GHENOVA and with the support of the following researchers: Víctor Blanco, Gabriel González, Miguel A. Pozo, and Justo Puerto.

WORKS CITED

Cuervas, F. J., Cabrera, A. J., Fontán, E., Brenes, P. J., Alejo, C., Tovar, A., Puerto, J., Conde, E., Ortega, F. A., & Hinojosa, Y. (2017). ARIADNA: Sistema automático de trazado de tuberías y canalizaciones en ingeniería. In: Ingenieros e industria L. Vilches Collado (Ed.), *Ingeniería Naval*, (No. 963, pp. 75–84). Asociación de Ingenieros Navales de España. http://www.ingenierosnavales.com/

Guirardello, R., & Swaney, R. E. (2005). Optimization of process plant layout with pipe routing. *Computers & Chemical Engineering*, 30(1), 99–114.

Min, J. G., Ruy, W. S., & Park, C. S. (2020). Faster pipe auto-routing using improved jump point search. *International Journal of Naval Architecture and Ocean Engineering*, 12, 596–604.

Park, J. H., & Storch, R. L. (2002). Pipe-routing algorithm development: Case study of a ship engine room design. *Expert Systems with Applications, 23*(3), 299–309.

Singh, J., & Cheng, J. C. P. (2021). Automating the generation of 3D multiple pipe layout design using BIM and heuristic search methods. In E. Toledo Santos & S. Scheer (Eds.), *Proceedings of the 18th international conference on computing in civil and building engineering (ICCCBE 2020). Lecture notes in civil engineering*, Vol. 98 (pp. 54–72). Springer. https://doi.org/10.1007/978-3-030-51295-8_6

4.4 DIGITAL FACTORY FOR SMALL- AND MEDIUM-SIZED ADVANCED TRANSPORT COMPANIES

Miguel Torres García and Gonzalo Quirosa Jiménez

ABSTRACT

The project develops the concept and implementation of Industry 4.0 for small- and medium-sized companies, which is currently lacking in the industrial sector. The aim is to obtain a methodology or procedure to facilitate the conversion of medium-sized industrial manufacturing companies into "digital factory" working models, in accordance with Industry 4.0.

INTRODUCTION

One of the missions that society entrusts to the university is the transfer of knowledge. The activities in this area include finding answers to the challenges and the needs of the industry. In relation to the above, the objective of this project, called FADIN 4.0, is to help small and medium-sized enterprises (SMEs) in the advanced transport sectors (automotive, naval, aerospace) and their supply chain, machining companies, and their supplementary to integrate into the new concept of digital factory (or Industry 4.0 paradigm). All this is intended to be achieved through the development of the digital factory concept for this type of companies through the transfer of knowledge and the use of existing limited R&D infrastructures.

STATE OF THE ART

The concept of Industry 4.0 is relatively recent and refers to the Fourth Industrial Revolution, which consists of a new way of organizing the means of production and the introduction of digital technologies in the industry. The main technological premises on which this concept is based are Internet of Things, cyber-physical systems, the "do it yourself" culture, and Factory 4.0. This includes, for example, the robotization of processes (in the Third Industrial Revolution, process automation was a general concept), the use of advanced automation strategies, advanced manufacturing, etc.

As an example of the importance of this industrial revolution, the European Commission has invested €500 million to define a series of Digital Innovation Hubs (DIH) to act as connectors and enablers for European industry to benefit from digital innovations by upgrading products and processes and adapting their business models to digital change (Moreno & García-Álvarez, 2018).

At present, many of the basic principles postulated by Industry 4.0 have yet to be implemented in practice. In the manufacturing industry, which is so conservative when it comes to introducing technological improvements in its processes, the establishment of complete connectivity of all factory equipment (Internet of Things concept) and the development of systems for capturing, processing, and mass analysis of data from this equipment to assist in decision-making (Big Data concept) are still in their infancy.

In this new paradigm, a series of "digital enablers" have been defined, which are the set of technologies that make it possible for this new industry to exploit its full potential. In effect, these allow the hybridization between the physical and digital worlds, such as linking the physical and virtual worlds to make the industry a smart industry.

On the other hand, the European Digital Agenda (Moreno & García-Álvarez, 2018) in its analysis of the industry defines the current context with key data, such as the fact that the most digitized companies are large companies (54%) compared to 17% of SMEs.

Most likely, the sectors to lead this transition will be aerospace, defense, industrial production, and automotive, where people already work alongside smart machines. However, the digitization of industry, or Industry 4.0, encompasses much more than technology. Companies must be prepared to undergo radical changes due to various factors, such as the speed of mass production, the volume and unpredictability of production, the increased fragmentation and reorientation of value chains, the importance of the customer in the production process, and the new relationships between research institutes, higher education and the private sector. All this transition is even more accentuated in SMEs, due to the fact that they neither have the appropriate means and resources nor specialized staff, and, therefore, they face a greater challenge than large companies that do have these resources, which are also specially adapted to them.

However, although the global, national, and regional industries are aware of the need to implement Industry 4.0 technologies in the market to maintain and strengthen its competitiveness, the reality is that there are still no solid solutions that integrate all the technologies of Industry 4.0, so as to respond to many of the challenges posed by the transformation of conventional factories to the concept of smart factories. This project meets this need for the development and integration of solutions adapted to industrial SMEs through a methodology that supports these companies in their process of hybridization of the physical and digital world, developing a technological platform adapted to the needs of industrial SMEs, setting up two demonstration centers that simulate a real production environment, preserving laboratory attributes that allow modifications at the specimen level, enhancing synergies between technologies, and developing and promoting training programs for those responsible for the digital factory of industrial SMEs, since the key lies in "transformation". This means that it is not just a matter of incorporating the latest technology but of transforming the business models of these companies through technologies.

CONTRIBUTION

The concepts of Industry 4.0, and all the technology associated with them, originate from their existence as part of a solution to a social and environmental problem. The strategic positioning of the market in the midst of digital factories is aligned with those trends that can have a more significant impact on its business, such as personalization, sustainability, or similar.

The main characteristics of the production model of Industry 4.0 or Digital Factory are flexibility, reconfigurability, and digitalization (Pallarés Martínez, 2018). However, there are two other transversal characteristics that are very relevant to the entire production model based on Industry 4.0 and are closely aligned with sustainability in its economic, social, and environmental aspects:

- People-centered model, regardless of the level of automation.
- Efficient model, ensuring maximum value while using the minimum resources necessary.

FADIN 4.0 covers these characteristics. Mainly, it stands out for being in line with the sectoral and horizontal strategies and policies of the EU. The project contributes to the objectives of the Europe 2020 Strategy to promote smart, sustainable, and inclusive growth, supporting innovation efforts and improving the competitiveness of a strategic sector for Andalusia.

The project aims to improve the current capabilities of industrial SMEs through innovations and technological developments necessary for their digital transformation. Also, it pursues synergies with other programs such as H2020, Shift2Rail, or Clean Sky, which are related to cleaner and more efficient production, with the manufacture of more sustainable and environmentally friendly components and structures. In this way, it can contribute not only to the economic development of the regions but also to the sustainability of the manufacturing sector associated with advanced transport, as well as to the conservation of resources.

Moreover, FADIN 4.0 and its actions are aimed at enhancing innovation in line with the flagship initiative "Resource Efficient Europe" (European Commission, 2011), which supports the transition to an efficient and low-carbon economy for sustainable growth. Thanks to the implementation of technologies related to Industry 4.0 or Smart Manufacturing, processes will be carried out in a more efficient way, reducing waste production and minimizing raw material consumption. Innovation in this manufacturing sector associated with advanced transport (automotive, naval, aeronautics) directly generates great economic opportunities and improved productivity, reducing costs and increasing the competitiveness of industrial SMEs in the cross-border regions of Andalusia and the European Union.

The project tends to improve investment and innovation security for industrial SMEs and to ensure that all policies take into account resource efficiency in a balanced way. Also, it is planned to contribute, through innovation in industrial SMEs, to the improvement of capacities of Andalusian important transport companies, as well as to enhance employability and economic and social sustainability of areas where there is a high risk of depopulation.

The feasibility of the project is based on the cooperation between companies and research organizations, as well as the training and implementation of innovations from research centers applied to the needs of the market. Therefore, based on the future own resources of the project partners, the results obtained by FADIN 4.0 are planned to be sustainable in the future from the quadruple perspective of research, training, development of improved protocols, and creation of business services.

ACKNOWLEDGEMENTS

This project has been carried out in collaboration with NINGENIA Design and Automation and Andalusian Foundation for Aerospace Development (FADA).

WORKS CITED

European Commission. (2011). Roadmap to a resource efficient Europe. https://eur-lex.europa.eu/legal-content/EN/TXT/PDF/?uri=CELEX:52011DC0571&from=EN

Moreno, B., & García-Álvarez, M. T. (2018). Measuring the progress towards a resource-efficient European Union under the Europe 2020 strategy. *Journal of Cleaner Production, 170*(1), 991–1005.

Pallarés Martínez, V. (2018). *Implementación de la Industria 4.0 en PYMES del Sector Productivo.* Universitat Politècnica de València. [Master Thesis] http://hdl.handle.net/10251/142963

4.5 A NOVEL APPROACH TO RADIATION HARDNESS ASSURANCE FOR AEROSPACE APPLICATIONS BASED ON MACHINE LEARNING

Yolanda Morilla García and Pedro Martín-Holgado

ABSTRACT

The space radiation environment damage at the part level has a severe impact on the reliability of missions. A rigorous methodology, the so-called Radiation Hardness Assurance (RHA), is needed to ensure that this does not compromise the electronics functionality and performance during the space systems life. Although radiation testing is the most decisive way of studying the radiation degradation, the increasing use of Commercial-Off-The-Shelf (COTS) devices and the NewSpace challenges are pushing the need of finding new approaches to assess the risk associated with the radiation environments. This work attempts to evaluate new RHA strategies based on Machine Learning (ML) philosophy.

INTRODUCTION

Radiation environments and their effects are a critical concern for materials and systems used in aerospace. The radiation damage has a severe impact on onboard electronics; it can induce malfunction, temporary loss of operation, or even mission loss. Electronic components are affected, so different effects upsurge, such as total ionizing dose—TID, displacement damage—DD, or single-event effects (SEEs).

The RHA methodology (Campola & Pellish, 2019) covers the activities to ensure that onboard electronics work according to their design specifications after radiation exposure. For several reasons, radiation testing has played a crucial role in revealing and characterizing vulnerabilities in systems. In NewSpace, considering the commercial off-the-shelf or commercially available off-the-shelf (COTS) devices and the cost-schedule constraints, testing every part type is not an option. The use of historical data from previous tests is already an important part of the device usage evaluation. Nevertheless, in order to meet and justify that there is no need to perform radiation testing, engineers should look for data that are not always easy to find, access, or interpret.

We present the early stage of a new concept in the strategy for ensuring radiation hardness in electronics. ML algorithms will be evaluated in the prediction of electronic components' behavior under radiation.

STATE OF THE ART

McGovern and Wagstaff (2011) already highlighted ten years ago how space was an excellent opportunity for new applications in the field of ML because it would allow greater autonomy. Until then, only one onboard operational ML had been uploaded to a mission: a support vector machine on the EO-1 spacecraft. It provided an additional data product in real time for identifying events. Other developments were performed, but none was uploaded, mainly due to risk considerations and the difficulty of proving sufficient security. A major breakthrough consisted in evaluating the impact of the severe radiation environment from space on the reliability of the learning algorithms themselves, and it was a focus of attention. Several opportunities for ML in space were then revealed, such as image analysis and reinforcement learning.

More recent works (Mohan & Tejaswi, 2020) declare the key role of Artificial Intelligence (AI) and ML for space exploration, and how they help space missions to solve the human error, time, and cost issues. Nowadays, it is assumed that AI, and in particular ML, still has some way to go before it is used extensively for space applications, but there are lots of projects where they have already been implemented (European Space Agency, 2021). For instance, when analyzing massive amounts of Earth observation data (too voluminous for humans to analyze themselves) or spacecraft telemetry data, examples such as safe operation of large satellite constellations, ability to avoid collisions (space debris) or rovers that can circumvent obstacles by autonomously finding their way through "unknown" fields, can be found.

AI is, therefore, widely used for improving onboard capabilities and analysis technics. The degradation of components due to radiation is an important constraint which limits severely spacecraft's total lifetimes, in addition to extreme environments, cost, and consumption of finite resources. The question is how AI can improve the knowledge of this parametric degradation. An answer to this question could be based on the predictive analysis of archival data from previous radiation tests carried out on similar electronic components. The use of historical data from previous tests as an alternative to new testing is not modern, but that approach was usually considered quite unreliable due to part-to-part and lot-to-lot variability. Nevertheless, thanks to ML, it is expected that valuable information can be obtained combining the extracted data from other "equivalent components" already tested under radiation and the behavior of the devices under normal conditions.

CONTRIBUTION

The CNA—*Centro Nacional de Aceleradores*—is a joint center depending on the *Universidad de Sevilla*, the Spanish National Research Council (CSIC), and the Andalusia Regional Government (*Junta de Andalucía*). Currently, two of the installed particle accelerators and the gamma irradiation system are used to carry out irradiation testing. ALTER Technology TÜV NORD is the leader in the field of electronic components radiation testing. The long fruitful relationship between CNA and ALTER, participating in joint projects and sharing the facilities, provides an ideal setting to offer a comprehensive radiation laboratory. However, in the frame of work of the current project PRECEDER, this consortium poses a new challenge: provide a novel predictive tool to the radiation community. Having previous knowledge about the behavior of the different components under radiation is quite interesting in the design phase of any system, more especially for NewSpace.

Nowadays, the number of small satellites launched by commercial and governmental administrations is growing up. The advantage of reducing production costs goes against the price of using radiation-hardened devices, too high in comparison to the baseline approach, the use of COTS. Additionally, it is important to consider the time consuming and budget increasing due to the performance of radiation testing on these components, with many different technologies.

To meet and justify that there is no need for performing radiation testing, it is necessary to search available data, based on previous radiation tests. To facilitate this arduous and tedious work, our group has developed PRECEDER as a database in the form of a web application, which will be included within VirtualLab[1] by ALTER.

Collected data are the base for a more ambitious objective, the data analysis focused on ML in order to obtain information about electronics behavior under radiation. Different data sources have been explored coming from ESA, NASA, JPL, ALTER, and other private companies. The first critical step in any ML project is data cleaning and preparation (García et al., 2016). So, tens of thousands of reports are being previously treated in order to ensure the uniformity, accuracy, and consistency of data. To eliminate the variability of the reporting and the wide number of different report templates, we were forced to develop a quite complex database structure.

Each irradiation test report includes several electrical parameters which have been measured in different conditions. Currently, millions of data are available, but substantial growth is expected due to the increase in space industry in the near future. As a novelty, the PRECEDER application, in an interactive way, offers the possibility to obtain a summary of conclusions or a prediction about the most relevant parameters in the radiation level of interest. Continuously updated, it will be public access with user-friendly interface, shortening the amount of time and efforts to find the information.

Once all the information is properly classified, cleaned, and clustered, we proceed to use different data science methodologies applying ML techniques, looking for some algorithms that generate a greater degree of confidence. Achieving high accuracy requires a large amount of data that is sometimes difficult, expensive, or impractical to obtain. In this field, where the available data is not so extensive for every component and the physical meaning is critical, supervised learning will be necessary. Integrating human knowledge into Machine Learning can significantly reduce data requirements, increase reliability and robustness of ML, and build explainable ML systems. Under this assumption, the last decision of clustering data from one or more radiation test reports must lie in radiation engineers that give support to data scientists.

First, we have studied TID reports for different transistor families. In view of our initial observations, from clustering and statistical data analysis, it seems to lead up to obtain valuable information about parametric degradation of electronic components. However, probably it should be mandatory to perform standard electrical measurements of critical parameters. Some weaknesses, such as part-to-part variability, must be considered as possible limiting factors. Even so, any reduction in time and costs is always welcomed. In the future, this kind of treatment could be included in the RHA methodology for aerospace applications.

ACKNOWLEDGMENTS

This project has been carried out in collaboration with ALTER TECHNOLOGY TÜV NORD S.A.U. and with the support of the following researchers: Manuel Domínguez Álvarez, Amor Romero Maestre, Iván Illera Gómez, Yolanda

Jiménez de Luna, José de Martín Hernández, José J. González Luján, and Fernando Morilla García.

Note

1. https://virtuallab.altertechnology.com

WORKS CITED

Campola, M., & Pellish, J. (2019). Radiation hardness assurance: Evolving for new space. RADECS2019 Shortcourse. https://nepp.nasa.gov/docs/tasks/047-Radiation-Hardness-Assurance/NEPP-CP-2019-Campola-RADECS-Paper-RHA-TN72757.pdf

European Space Agency. (2021). Artificial Intelligence in space. https://www.esa.int/Enabling_Support/Preparing_for_the_Future/Discovery_and_Preparation/Artificial_intelligence_in_space

García, S., Ramírez-Gallego, S., Luengo, J., Benítez, J. M., & Herrera, F. (2016). Big data preprocessing: Methods and prospects. *Big Data Analytics*, *1*(9). https://doi.org/10.1186/s41044-016-0014-0

McGovern, A., & Wagstaff, K. L. (2011). Machine learning in space: Extending our reach. *Mach Learn*, 84, 335–340. https://doi.org/10.1007/s10994-011-5249-4

Mohan, J. P., & Tejaswi, N. (2020). A study on embedding the Artificial Intelligence and machine learning into space exploration and astronomy. In: Hemanth, D., Kumar, V., Malathi, S., Castillo, O., & Patrut, B. (Eds.), *Emerging trends in computing and expert technology. COMET 2019. Lecture notes on data engineering and communications technologies, vol 35*. Springer. https://doi.org/10.1007/978-3-030-32150-5_131

4.6 OPTICAL SENSING AND SELF-LEARNING APPROACH TO ESTIMATE THE STATE CONDITION OF RAILWAY INFRASTRUCTURE SUBLAYERS

Francisco A. García Benítez and Fernando Lazcano Alvarado

ABSTRACT

The objective pursued is the implementation of a technique for intensive information capture of the state of the deep layers of the infrastructure of transport linear works, in order to (a) detect faults, (b) predict the evolution of the state as a function of time, (c) estimate the necessary maintenance operations, and (d) plan the required interventions. All this is focused on achieving greater efficiency in the management of the conservation of this kind of infrastructure. The data obtained in real time correspond to the tests carried out on a substructure section model housed in a scaled sublayer's test-rig with installed Fiber Bragg Grating (FBG) optical sensors. Finally, the methodology is described, based on data analytics and Machine Learning techniques, in order to infer the severity of measured deformations and failures.

INTRODUCTION

A significative share of transport infrastructures is composed of linear assets, such as roads and rail tracks. The social and economic relevance of these constructions force the stakeholders to ensure a prolonged health/durability, but inevitable malfunctioning, breaking down, and out-of-service periods arise during their life cycle. Of all assets that make up an infrastructure, a very relevant asset is the set of sublayers. The failure or deterioration state condition of the deep layers absorbs high levels of economic resources and generates important social impacts.

Predictive techniques tend to diminish the appearance of unpredicted failures and the execution of needed corrective interventions. The use of historical asset conditions and their corresponding maintenance interventions data, and data analytical approaches facilitate the envisaging of the adequate maintenance interventions to be conducted before failures show up.

This paper presents (i) a technological approach to capture, in real time, the stress–strain state of the infrastructure deep sublayers using optical sensing based on FBG technology, (ii) an approach to infer the state condition of the layers using data analytics, and (iii) an automatic learning procedure based on self-learning rules from automatic learning from false positive/negatives.

STATE OF THE ART

The detection of maintenance alerts is generally based on the inspection of the state of the assets through the visualization/auscultation/measurement of the explanatory characteristics of the involved asset. The evolution of these characteristics, estimated quantitatively or qualitatively, using projection techniques (i.e., regression) or qualitative (e.g., experience), and cross-checking with thresholds and limits (defined by technical standards prescribed by either the administration or infrastructure regulator), has been the main tool to anticipate possible failures or deficiencies in the operation of assets.

For decades, the inference of deterministic or probabilistic models based on a priori explanatory characteristics (i.e., empirical-mechanistic models) has been the most pronounced trend (1960–1990). Currently, developments have focused on replacing this way of proceeding by data-based modeling, making use of data mining and Machine Learning techniques (1990–2015); this is based on the increasing availability of data captured from activities, auscultations, and monitoring campaigns.

In the field of transport infrastructures, since the 1990s, the trend has focused on predicting the evolution of the state of certain components using ML techniques and data analytics; however, in most cases, attention has been focused on the prediction of easily accessible active associated indices (pavement on highways, rails and ballast on railways) or to the global state of the infrastructure using data from external geometry (1995–2020). In relation to the estimation of maintenance interventions, there have also been important advances referenced in various publications on the state of the art on this subject.

In the field of monitoring the state of deep layers of the substructure, advances have been more limited due to the cost and difficulty of sensorization. This study advances the use of data-harvesting methodologies to capture the state of the substructure at deep levels based on optical sensorization techniques and its subsequent treatment to characterize the state of integrity.

These data, properly processed, constitute an important source of information to infer the state of the substructure. The application of some techniques developed by the research team, for the exploitation of data associated with the surface assets of road networks and railway lines (Morales et al., 2017, 2018, 2020; Reyes, 2018) is at the background of this communication.

CONTRIBUTION

The contribution of the research hovers on three pillars: (a) to build and sensorize a test-rig to generate and harvest data from infrastructure sublayers, (b) to exploit the harvested data to infer a cause-effect (loading-failure) relationship using data analytics and automatic learning techniques, (c) to apply the prediction methodology derived by the research group to estimate maintenance interventions.

a) *Testing Rig*

At this stage of the project, a first test-rig scaled 1:7 has been built to simulate the condition of the infrastructure sublayers of a railway track section. The section is made of several layers of sand and gravel with embedded thin monitoring metal plates (0.25 and 0.1 m²), between layers.

The monitoring plates house strain and temperature sensing (10 sensors each) based on Bragg Grating sensing (FBG). An eternal read-out system (interrogator), connected to the plates using silica-glass fibers, injects optical signals of modulated laser pulse-lights that interact with the sensors and that are altered by the strain suffered by the sensors when some perturbation occurs.

The FBG sensing system is characterized by its high sensitivity to any alteration either on the loadings, experienced by the track model on the infrastructure surface, or on faults appearing inside the layers.

In order to quantify the sensitivity of the substructure model, due to internal failures, a set of air pressurized cushions (with externally controlled pressure) are set inside some of the layers to artificially generate faults under several conditions.

By crossing out the level of interaction detected by the interrogator system, spatial mapping of the strain distribution inside the layers is generated and translated to soil displacements (i.e., breaks, glides, settlements). Similar tests are performed under variable loadings affecting the track surface in order to characterize the FBG sensing capability.

A second test-rig 1:1 is in process of construction, which holds a full real track section (rail, ties, ballast, sub-ballast, sublayers), with a more extensive sensorization level.

b) *Data Generation and Exploitation*

A multiplicity of tests is performed in order to generate a rich database. The tests should always be performed under variable settings (i.e., surface loadings, layers faults). The distributed strain measurements of the sensor chain array, provided by the interrogator, have to be post-processed first to map data to a free-fault condition of the layers. The second step consists in correlating/matching sensor data with the loadings and faults. The third step applies a set of Machine Learning methodologies to infer an estimating frame cause-effect. Details about this last step can be found in the references (Morales et al., 2018, 2020).

ACKNOWLEDGMENTS

This project has been carried out in collaboration with Grupo, Azvi, and with the support of Francisco J. Morales Sánchez (PhD senior researcher) and Antonio Valverde Martin (technician).

WORKS CITED

Morales, F. J., Reyes, A., Caceres, N., Romero, L. M., & Benitez, F. G. (2018). Automatic prediction of maintenance intervention types in roads using machine learning and historical records. *Transportation Research Record: Journal of the Transportation Research Board*, 1–12. https://doi.org/10.1177/0361198118790624

Morales, F. J., Reyes, A., Caceres, N., Romero, L. M., Benitez, F. G., Morgado, J., & Duarte, E. (2020). Machine learning methodology to predict alerts and maintenance interventions in roads. *Roads Materials and Pavement Design*, *21*. https://doi.org/10.1080/14680629.2020.1753098

Morales, F. J., Reyes, A., Caceres, N., Romero, L. M., Benitez, F. G., Morgado, J., Duarte, E., & Martins, T. (2017). Historical maintenance relevant information roadmap for a self-learning maintenance prediction procedural approach. *Materials Science and Engineering Conference Series, 236.* https://doi.org/10.1088/1757-899X/236/1/012107

Reyes, A., Morales, F. J., Caceres, N., Romero, L. M., Benitez, F. G., Morgado, J., Duarte, E., & Martins, T. (2018). Automatic prediction of maintenance intervention types in transport linear infrastructures using machine learning. In *Proceedings of 7th transport research arena TRA2018.* Paper ID 10392. Vienna, Austria, 16–19 April 2018.

4.7 POWER-EFFICIENT ANALOG-TO-INFORMATION IMAGE SENSORS: THE FUEL OF AI MICROSYSTEMS

Ángel Rodríguez-Vázquez and Juan A. Leñero-Bardallo

ABSTRACT

This project explores innovative architectural concepts for image sensors with embedded intelligence. The main attributes of these architectures are the use of non-Von Neumann paradigms and the extensive usage of event-driven concepts. The goal is to achieve vision with unprecedented throughput and energy efficiency.

INTRODUCTION

Miniaturized, *intelligent* microsystems are at the central focus of semiconductor roadmaps and strategic nanoelectronic agendas (International Roadmap for Devices and Systems, 2020). They are enabled by the convergence of *heterogeneous* technologies and targeted to give electronic systems the ability to interact with the cyber-universe of analog information carrier signals. Besides being intelligent, in the sense of capable of *adapting* their response to arbitrary stimuli and unstructured environments, these systems must *communicate*, have a minimal *size* and *weight*, minimum *energy* consumption, and, if possible, energy *autonomy*. The target is to support human–machine interaction with electronic systems and appliances, have these systems connected, construct robots capable of detecting subtle surrounding changes and reacting rapidly, allow vehicles to identify risky situations, escape from them, etc. We also want electronic systems to resemble natural beings' outstanding capabilities regarding sensory data acquisition, analysis, interpretation, and reaction thereof. Challenges like these shape the domain of intelligent microsystems, a domain where the handling of information conveyed by light energy and the sense of *vision* is expected to play a significant role. Lately, the vision sense handles around 50% of the information that humans interchange with the environment.

STATE OF THE ART

Computer vision is one of the main areas within the ecosystem of intelligent sensory systems. Strategic agendas coincide when pointing out the widespread deployment of miniaturized computer vision systems as one of the main engines of R&D-and-Innovation in IT and nanoelectronics. The challenge is to conceive SWaP-optimized *artificial vision architectures* with embedded analysis capabilities and large efficiency in the number of operations completed per unit of time and energy. However, conventional processing architectures cannot meet these challenges. These traditional architectures adhere to the paradigm of encoding the analog information in the digital domain (using *digital imagers*) and relying on Von Neumann's digital processing architectures for analysis (A. Rodríguez-Vázquez et al., 2018). It implies encoding, storing, and transmitting the whole set of *raw* image data, following a *frame-by-frame* operation basis that uses enormous processing and energy resources while handling meaningless data. This approach significantly differs from that of *natural vision systems* as they operate by extracting *information*, instead of just raw data, right from the very front ends, at the edge of the information processing chain. Retinas, the front ends of natural vision systems, are *information-centric* front-end devices, while imagers used in conventional computer vision systems are *data-centric* devices.

The TIC179 research group and the company Teledyne-Anafocus (which was launched in 2001 by members of the TIC179 with the name Innovaciones Microelectrónicas S.L.) have conducted R&D on bio-inspired sensor front ends and smart imagers for decades. Contributions follow tracks that are concurrently explored by many other companies and academic institutions worldwide, namely:

- By incorporating *early-vision* processing tasks right in the focal plane, they are completed concurrently to light capture. The aim is to encode and communicate *features*, instead of frames. The approach uses a parallel processing paradigm alternative to Von Neumann's architectures but conferred with the *software-programming* capabilities of such a paradigm (Rodríguez-Vázquez et al., 2018).

- By using information encoding techniques similar to those employed by natural vision systems, including (i) the detection of spatial-temporal contrasts within visual scenes, (ii) the usage of pixels that trigger when they detect transient lighting changes, and (iii) use of peak luminance sensors that perform a light-to-frequency conversion (Leñero-Bardallo et al., 2018). Similar to feature-extraction front ends, all these techniques achieve large power efficiency and bandwidth.
- By exploring the ultimate physical limits of light carriers through detectors able to count individual photons and signaling the photon arrival events in time.

CONTRIBUTION

They are aligned to the challenge of improving the energy efficiency and performance of smart-imaging and vision sensor chips. These chips are lately the fuel of efficient sensor interfaces and hence crucial to incorporate Artificial Intelligence at systems targeted for the following AI technologies: (T1) IoT, (T2) robotics, and (T3) augmented/virtual reality. Targeted applications are, among others, in the following sectors: (S1) advanced industry linked to transportation, (S2) health and social welfare, and (S3) mobility and logistics.

Specific contributions of this project are summarized in the points below.

- The conception of the pixel and the chip architecture of a high dynamic range asynchronous image sensor with embedded solar energy harvesting capability. Its photodiodes operate in the reverse region of operation when measuring the lighting levels to which they are exposed. Also, they automatically switch to the photovoltaic regime to collect energy that feeds the sensor itself. Innovative circuit techniques allow these pixels to operate with down to 0.35 V voltage levels, compatible with the operating voltages of photodiodes in the photovoltaic region (Gómez-Merchán et al., 2020).
- The conception of two asynchronous solar sensors for vehicle navigation based on key-points tracking. They operate at high speed and employ photodiodes in the photovoltaic region of operation to reduce energy consumption. Only pixels that receive solar radiation generate output data. The first one measures the illumination

levels of the illuminated pixels to determine the sun's position by processing its outputs. The second provides directly as an output the coordinates of the centroid of the illuminated region, representing a great advance compared to the state of the art.

- The conception of new families of pixels and sensors capable of single-photon detection (through Singe Photon Avalanche Diodes) and embedded intelligence at pixel and sensor levels. These include distributed photon counters and distributed Time-to-Digital converters to capture 2D/3D scenes. They are targeted to support a new generation of solid-state LiDAR systems, on the one hand, and for the implementation of micro Positron Emission Tomography systems, on the other hand (Vornicu et al., 2020).

All architectural ideas explored in the project have been demonstrated through dedicated CMOS chips and camera modules built with them.

ACKNOWLEDGMENTS

This project has been carried out in collaboration with TELEDYNE-ANAFOCUS and with the support of the following researchers: Ana Gónzalez and José Ángel Segovia de la Torre.

WORKS CITED

Gómez-Merchán, R., et al. (2020). A comparative study of stacked-diode configurations operating in the photovoltaic region. *IEEE Sensors Journal, 20*(16), 9105–9113. https://doi.org/10.1109/JSEN.2020.2987393

International Roadmap for Devices and Systems, IEEE, Edition 2020. https://irds.ieee.org/

Rodríguez-Vázquez, A., et al. (2018). CMOS vision sensors: Embedding computer vision at imaging front-ends. *IEEE Circuits and Systems Magazine, 18*(2), 90–107. https://doi.org/10.1109/MCAS.2018.2821772

Vornicu, I., Carmona-Galán, R., & Rodríguez-Vázquez, A. (2020). CMOS SPAD sensors with embedded smartness. *International SPAD-Sensor Workshop.* Edinburgh, Scotland (UK), June 2020.

Endogenous Land-Based Resources, Agroindustry, and Tourism

5

This chapter is of vital importance for the region of Andalusia, due to its surface area (it is similar to some European countries), the importance of agriculture as a result of its climatic conditions (it is considered the pantry of Europe due to the number of fruits and vegetables that are produced in Andalusia and consumed in other countries), and tourism reflected in the large number of tourists that the region receives annually, attracted by its historical heritage, the sun, and the beach. These three pillars, due to their historical tradition and economic impact, become optimal fields to incorporate innovation using Artificial Intelligence techniques. The productivity and efficiency of these products and services should mark the roadmap for the coming years, consolidating existing industries, as well as emerging new ones in the market niches that arise with the digital transformation.

The chapter presents case studies such as the planning of maintenance activities in water distribution networks in urban centers, using Machine Learning techniques, to reduce pipe substitutions and prioritize heritage management that minimizes the impact of incidences of breakages. In addition, another work on the use of geographic information with topology is shown, using Machine Learning and Big Data techniques and integrating them into special geo-visualization platforms with large volumes of environmental data for future use in the environmental and socioeconomic impact.

Regarding the agricultural sector, a case is shown with real-time monitoring of super-intensive olive trees to improve quality and yield, using deep learning techniques. Olive oil is the main crop in Andalusia and its modernization

DOI: 10.1201/9781003276609-5

with analytical techniques will allow maintaining a competitive product compared to other producing countries with low labor costs. Likewise, advances in cyber-physical systems are presented, which help transform agriculture into intelligent activities and with greater control of the state of crops, particularized in olive groves.

Finally, the use of digital spatial techniques in the knowledge and conservation of archaeological sites and historical architecture is presented. A melting pot of civilizations has passed through the Andalusia region: Phoenicians, Romans, Arabs, and the impact of the discovery of America, and all of them have left their mark on a region rich in heritage, which has to be maintained with quantitative techniques such as building information modeling (BIM) and Geographic Information Systems (GIS).

5.1 PRIORITIZING WATER DISTRIBUTION AND SEWER NETWORK MAINTENANCE ACTIVITIES

Luis Onieva Giménez and Alicia Robles-Velasco

ABSTRACT

Asset management in water supply and sewerage infrastructures seeks correct planning and prioritization of the maintenance activities on their network elements. To determine the replacement priority for each network asset, an original priority model is utilized. It is based on a risk index that integrates both the probability and consequences of failure. Besides, when designing and performing work programs, water companies generally abide by the street's topography and other urban elements, such as complete streets or street sections. This work considers street sections between the two nearest intersections as the operational unit.

INTRODUCTION

The long-term sustainability of hydraulic infrastructures is based on a correct and effective maintenance strategy. Establishing the right time to replace a water distribution and sewer network element is essential to properly manage the infrastructure. In this regard, an adequate definition of the replacement needs of each network element is key to the right infrastructure management. Also, this process depends on the reliable condition and criticality information of every asset, for which an integrated, complete, and up-to-date data system is fundamental.

Moreover, this replacement priority has been usually computed for individual water and sewer pipe objects. Nonetheless, when designing intervention programs, the use of this hydraulic element constitutes an ideal and abstract approach. As a matter of fact, water utilities do not consider pipelines to plan interventions but perform work programs according to the street's topography. The inclusion of geographical and urban criteria into the planning process helps minimize the affection on the traffic flow and the neighborhood, as well as fosters coordination with other public infrastructure projects.

STATE OF THE ART

The reviewed literature in this field addresses methodologies to develop and implement different infrastructure asset management strategies. More specifically, the reviewed works target the two following aspects: the utilization of a priority model to determine the replacement urgency of the network assets and the consideration of infrastructure 'units' that represent the smallest replacement element and abide by the street's topography or urban factors.

For example, Elsawah et al. (2016) define corridors as the street section between the two nearest intersections. When intervening a corridor, all the water, sewerage, and road infrastructure assets within that segment are simultaneously replaced. Also, the authors developed an overall risk index for the whole corridor segment, integrating the replacement needs of the three coexisting infrastructures. This methodology is applied to a borough of the City of Montreal (Canada).

A similar example can be found in Tscheikner-Gratl et al. (2016), where entire street sections are ranked for rehabilitation. These elements correspond to segments between valves or manholes and include the road, water supply, and sewerage infrastructures. Also, to avoid very long sections, a certain length threshold is set. Again, an overall priority index for the whole section is calculated. A case study of an Austrian city with 130,000 inhabitants is presented.

Lastly, Shahata and Zayed (2016) developed another risk assessment model for road, water distribution, and sewer networks. This model was implemented in the City of Guelph (Canada).

From the reviewed literature, it can be drawn that the utilization of pipe objects as operational units is not practical to develop water replacement strategies. Thus, the definition of a workable intervention unit complying with urban elements, such as street sections, is essential.

Then, the replacement needs for a given corridor or street segment have to be established. The reviewed works propose an overall criticality index that is computed on the basis of the previous calculation of an index for each infrastructure separately. This approach can lead to some difficulties when setting the relative importance of each infrastructure. The utilization of an index that addresses a section as an integrated and indivisible unit, whose replacement priority does not require the definition of the relative importance between the diverse infrastructures is still missing in the literature. Our contribution to the literature in this regard is described in the following section.

CONTRIBUTION

Our main contribution has been to establish the replacement needs for every network operational unit and rank them for intervention. In the first place, the operational unit had to be defined. Thus, we proposed complete streets or street sections between intersections to be the smallest replacement element.

Next, an original procedure to determine the replacement needs for network assets and to rank them for intervention was required. To this end, a risk index (RI) was developed (Muñuzuri et al., 2020). This index is based on the probability and consequences of failure of a given pipe.

First, the likelihood of failure was obtained through AI technologies. More specifically, the Machine Learning techniques *logistic regression* and *support vector machine* were utilized. They serve as a predictive system, since they produce an output variable that can be used as a failure probability. Eight variables were considered to explain pipe failures: material, diameter, age, length, number of connections, network type, pressure fluctuation, and previous record of failures. The results obtained show that the number of unexpected failures could be significantly reduced. In fact, with either of both techniques, and according to the historical data from the city of Seville (where this methodology was applied), around 30% of failures could have been avoided by replacing only 3% of the network's pipes (Robles-Velasco et al., 2020).

On the other hand, the factors measuring the consequences of failure are the water leakage flow and demand for supply pipes, the maximum evacuation flow for sewer pipes, and the pipe criticality.

The RI can be easily assigned to every pipe object, providing a replacement priority for all of them. However, pipe objects do not represent the operational unit. The RI has to be computed at a "street section" level. Thus, the probability and consequences of failure of the underlying pipes have to be transferred to their corresponding street sections. To this aim, the following procedure was implemented.

First of all, it has to be noted that street sections and infrastructure objects are linked by geographical superposition: every network pipe is related to the section it runs through. So, the failure probability of a street section is computed as the average probability of failure of every pipe included in that section. This average value is weighted by the inner pipe's length. Since pipes may belong to more than one street section, only the length of the pipe inside the given section is considered. Afterward, the leakage flow, demand, and maximum evacuation flow for each section correspond to the maximum value of all their constituting pipes. Also, a section is considered relevant or critical when at least one of their inner pipes is, indeed, critical. Once every section is characterized by these five factors, its RI and intervention priority are obtained.

Finally, this methodology was fully integrated into an easy-to-use software application. It automatically processes the input data, such as the network's condition and criticality, and computes a RI value to every street section. This user-friendly tool also allows setting the planning horizon and the long-term objective and generates a prioritized list of street sections to be intervened. The development of this software was carried out by the aggregated partner GUADALTEL S.A.

The here presented methodology was applied to the water distribution and sewer networks of Seville (Spain), whose complete infrastructure system has a total length of approximately 7,000 km, combining both the supply and sewer networks.

This approach can help establish a replacement priority for every network asset and rank them for intervention, which is the basis to guarantee the long-term sustainability of the infrastructure. Besides, since this approach is based on a risk model, combining both probability and consequences of failure, it aids to target those assets whose preventive replacement could avoid much higher future economic and social costs.

Furthermore, this approach considers street sections as the smallest replacement units, which may be advantageous for various reasons. First, given that an intervention generally affects single streets, the impact on pedestrians or the traffic flow is reduced. Also, it benefits the possible coordination of the hydraulic network maintenance tasks with other infrastructures coexisting projects, which can help avoid duplicated road closures and the possible undermining of the company image.

Therefore, thanks to the integrated risk-based priority model and the utilization of workable and functional operational units, this methodology can be seen as a useful and practical decision support system to prioritize maintenance tasks and efficiently invest to guarantee the long-term sustainability of the infrastructure.

ACKNOWLEDGMENTS

This project has been carried out in collaboration with GUADALTEL, S.A. and with the support of the following researchers: Elena Barbadilla Martín, Alejandro Escudero Santana, María Rodríguez Palero, and Alicia Robles Velasco y Cristóbal Ramos Salgado.

WORKS CITED

Elsawah, H., Bakry, I., & Moselhi, O. (2016). Decision support model for integrated risk assessment and prioritization of intervention plans of municipal infrastructure. *Journal of Pipeline Systems Engineering and Practice, 7*(4). https://doi.org /10.1061/(ASCE)PS.1949-1204.0000245

Muñuzuri, J., Ramos, C., Vázquez, A., & Onieva, L. (2020). Use of discrete choice to calibrate a combined distribution and sewer pipe replacement model. *Urban Water Journal, 17*(2), 100–108. https://doi.org/10.1080/1573062X.2020.1748205

Robles-Velasco, A., Cortés, P., Muñuzuri, J., & Onieva, L. (2020). Prediction of pipe failures in water supply networks using logistic regression and support vector classification. *Reliability Engineering and System Safety, 196*(October 2019), 106754. https://doi.org/10.1016/j.ress.2019.106754

Shahata, K., & Zayed, T. (2016). Integrated risk-assessment framework for municipal infrastructure. *Journal of Construction Engineering and Management, 142*(1). https://doi.org/10.1061/(ASCE)CO.1943-7862.0001028

Tscheikner-Gratl, F., Sitzenfrei, R., Rauch, W., & Kleidorfer, M. (2016). Integrated rehabilitation planning of urban infrastructure systems using a street section priority model. *Urban Water Journal, 13*(1), 28–40. https://doi.org/10.1080 /1573062X.2015.1057174

5.2 NEW DATA STRUCTURE (ASYMMETRICAL AND MULTISCALE GRID) FOR GEODATA INTEGRATION AND INPUT FOR MACHINE LEARNING ANALYSIS

Juan Pedro Pérez Alcántara and Esperanza Sánchez Rodríguez

ABSTRACT

Geographic data integration has been an ongoing problem when trying to approach multivariate analysis. Besides, workflows for new data analysis techniques and technologies in the Machine Learning and Big Data parallel computing domains remain a challenge for geographic data because of the intrinsic links between features due to topological relationships. Heterogeneous data structures, dissymmetry in data induced by the scale, and the geometric nature of geographic information, especially when using vector structures, hinder multivariate analysis. To overcome these difficulties, a new data structure, asymmetrical multiscale grids, and a companion software framework to build and manage them are proposed.

INTRODUCTION

Heterogeneous environmental and socio-demographic geographic data integration (Ojeda Zújar et al., 2021; Batista e Silva et al., 2013) is a vital process for multivariate spatial analysis. Different geometry types, like raster and vector and, within the latter, the use of polygon, line, and point features, coupled with intrinsic geoinformation traits like scale significance, pose a challenge to this critical pre-analytical step. Moreover, geodata, thanks to its tightly interrelated nature due to the underlying topology, lacks a common framework when it comes to mapping the data to implement parallel processing workflows. To provide this framework, a new data structure, called asymmetrical multiscalar grids, is proposed.

To test this framework, an array of big data sets has been chosen. These data sets have been converted into the new format to test several solutions: statistical methods to transform and integrate the data, parallel computation of this integration in a cloud environment, and suitability of the final product to feed Machine Learning algorithms. The area of testing is the Autonomous Community of Andalusia, southern Spain, with 83,000 km². Data sets cover demographics, cadastre, household income, and several environmental variables. A cloud microservice application has been developed to integrate the data in a parallel computing environment and to serve data via an Application Programming Interface (API) service. This API is used to power several web visualization solutions.

STATE OF THE ART

With the increasing production of semi-automatic geographic data, especially in the earth sciences area, where computational models produce vast quantities of data sets on a regular basis, integration of heterogeneous data sources has become a matter of research in workflows, standards, and technologies (Álvarez Francoso, 2016). Global spherical hexagonal grids have been the subject of most of these works; among them, the discrete global grid system (DGGS) standard of the Open Geospatial Consortium (2017) (Bondaruk et al, 2019; Birch et al, 2007) presents a "hierarchical sequences of equal area tessellations on the surface of the Earth". By relying on this common framework, thematic specialists can provide the results of their research in a common, discretized reference, compatible with the outputs of other analyses. This common reference allows for standardization not only of procedures on the grid but of in homogeneous data that can be used more easily by non-specialists as input for multivariate analysis.

CONTRIBUTION

The result of this project is a software platform that produces data in a new geographic information format called the asymmetrical multiscalar grid. This format discretizes the geographic space in a manner that resembles a raster data structure. However, the grid is not pre-generated but mathematically defined so each cell can be computed in real time. This grid is used to integrate geodata sources by a variety of statistical means, depending on the nature of the source data (vector or raster, and the geometry of the former). Only cells that have colliding source data receive a data, being this data only the one that affects the given cell. Therefore, the grid is asymmetrical in the two dimensions of geographic data: alphanumeric and geometric. Cells without data don't physically exist in the final output, and each cell contains only the data set that collides with it. This way, there can be cells with a thousand variables while others may have only a dozen, hence the asymmetry in the alphanumeric data. This is achieved by capitalizing on nonstructured data formats commonly found in NoSQL databases, like for example the JSON format. Each cell ends up containing a thematic vector composed of a certain number of variables. Empty cells don't even exist in the final output.

On this data structure, microservice architecture has been developed to integrate original data into the new format and to serve its data in several forms. This service API provides data to remote clients. The data stored in the vectorized alphanumeric variable reservoirs of each cell can be explored and filtered, obtaining multivariate vectors that are the common input format for most of the Machine Learning (ML) algorithms. To test the suitability of this format for ML, both an unsupervised and a supervised method have been integrated into the platform. For unsupervised processing, the K-Means algorithm is used, and for the supervised, Random Forest (Bressert, 2012). The platform can apply these algorithms successfully to the integrated data stored in cells, proving that it is useful for this kind of application.

To visualize the data, some web client technologies have been used. One is a custom design that capitalizes the mathematical definition of the grid, therefore minimizing the transit of data between client and server, because the geometries are generated on the fly in the web browser by the geovisor. Given that the geovisor is receiving the raw data contained in the cell's vector in JSON format, it can use this information for a variety of mixed graphical data exploration techniques, combining for example statistical graphs like histograms with highly interactive maps. These two techniques, combined and in synch, can produce highly flexible and expressive means for data exploration. This geovisor leverages the latest standard stack of web technologies, such as HTML5 and WebGL (Belmonte, 2015; Anthes, 2012), along with libraries like D3, to enable rich visualization clients. Another prototype has been created by exporting data into the CARTO platform, which is also a tool specialized in geographic data exploration.

ACKNOWLEDGMENTS

This project has been carried out in collaboration with Geografía Aplicada S.L. "GEOGRAPHICA" (CARTO) and with the support of the following researchers: José Ojeda Zújar, Ismael Vallejo Villalta, Juan M. Camarillo Naranjo,

Jesús Mª Rodríguez Leal, Esperanza Sánchez Rodríguez, Natalia Limones Rodríguez, and Juan Pedro Pérez Alcántara.

WORKS CITED

Álvarez Francoso, J. I. (2016). *Geovisualización de grandes volúmenes de datos ambientales. Diseño e implementación de un sistema para el acceso y la difusión de datos globales.* PhD Thesis. Universidad de Sevilla.

Anthes, G. (2012). HTML5 leads a web revolution. *Communications of the ACM*, 55(7), 16–17.

Batista e Silva, F., Gallego, J., & Lavalle, C. (2013). A high-resolution population grid map for Europe. *Journal of Maps*, 9(1), 16–28.

Belmonte, N. (2015). Data visualization techniques with WebGL. *WebGL Insights*, 297, 297–315.

Birch, C., Oom, S. P., & Beecham, J. A. (2007). Rectangular and hexagonal grids used for observation, experiment and simulation in ecology. *Ecological Modelling*, 206(3–4), 347–359.

Bondaruk, B., Roberts, S. A., & Robertson, C. (2019). Discrete global grid systems: Operational capability of the current state of the art. *Spatial Knowledge and Information—Canada—Proceedings, 7*(6), 1–9.

Bressert, E. (2012). *SciPy and NumPy*. O'Reilly.

Ojeda Zujar, J., Paneque Salgado, P., Sanchez Rodriguez, E., & Perez Alcantara, J. P. (2021). Geografía de la renta de los hogares en España a nivel municipal: Nuevos datos y nuevas posibilidades de geovisualización, exploración y análisis espacial en entornos *cloud*. *Investigaciones Geográficas, 76*, 9–30.

5.3 DEVELOPMENT OF YIELD MONITORING SYSTEM BASED ON COMPUTER VISION FOR SUPER-HIGH-DENSITY OLIVE ORCHARD

Orly Enrique Apolo-Apolo and Manuel Pérez-Ruiz

ABSTRACT

New super-high-density (SHD) olive orchards designed for mechanical harvesting are becoming increasingly common around the world. However, precision farming techniques such as yield monitoring and mapping are still a challenge. This study presents a new approach based on deep learning algorithms to continuously monitor yields during the harvesting operation. A pretrained model (Fast R-CNN) was implemented for the detection of two classes: *olives* and *residues*. Olive fruit detection was used to automatically count the number and determine the percentage of residue harvested. Detection accuracy with values of 80% was obtained with this novel model in both classifications.

INTRODUCTION

Yield monitoring and mapping during the harvesting process are essential for farmers, since they help them gain access to detailed representations of their crop fields and make olive orchards more profitable. However, unlike other crops such as cereals where these techniques are already mature, in SHD orchards yield monitoring systems are still an unfilled gap. Probably because mechanically harvested SHD olive tree plantations have been established in the last decades in Spain (Pérez-Ruiz et al., 2018), which requires time to implement innovations related to harvesting machinery.

On the other hand, the development of Artificial Intelligence (AI) algorithms, particularly the ones based on deep learning techniques, have shown transformative impact in the agriculture sector (Kamilaris & Prenafeta-Boldú, 2018), providing successful results in agricultural robotics, crop health, yield estimations, growth progress detection, ripeness, and fruit size detection among others. Moreover, the emergence of ISOBUS technology standards has allowed communication and data transfer between tractors and agricultural machinery. Hence, the aim of this research is to develop a machine vision yield monitor for a super-high-density olive orchard suitable to be mounted on a mechanical olive harvester.

STATE OF THE ART

From the point of view of the agricultural, few efforts have been done in developing solutions for yield monitoring and mapping using AI. A machine vision-based citrus fruit counting system was developed for a continuous canopy shake and catch harvester was proposed by Chinchuluun et al. (2009). Colmenero-Martínez et al. (2018) propose an automatic trunk-detection system for intensive olive harvesting. On the other hand, a machine vision yield monitor for the counting and quality mapping of shallots was developed by Boatswain Jacques (2021). The most relevant achievement has been suggested by Bazame et al. (2021). They implemented a computer vision algorithm to detect and classify coffee fruits and map the fruits' maturation stage during the harvesting process. This study is significant because the machinery used in SHD olive orchards and the one used for harvesting coffee are similar. As a result of their research, accuracy greater than 80% in detecting coffee fruits is achieved.

According to the literature review, the most important components of yield monitoring and mapping are the global positional system receiver (GNSS), mass flow sensor, and yield monitor (Chung et al., 2016). The combination of all these technologies provided one of the most valuable sources of spatial data for precision agriculture, the yield maps (Ahmad & Mahdi, 2018). Traditionally, yield monitoring and mapping systems were based on volume or weight measurements. However, with advancements in new technologies, especially computer processing speed and AI, the door has been opened to more precise developed solutions.

CONTRIBUTION

An image classification model based on object detection was trained to recognize two classes: *olives* and *residues*. Both classes were included because in many cases the olive buyers reduce the cost of the crop production depending on the percentage of residues detected when the olives arrive at the factory. For the image acquisition, a small conveyor belt prototype was designed as the conveyor found on the straddle harvester. A tiny tank to store a known volume of olives was attached at the beginning of the conveyor belt. Conveyor belt speed was regulated using an electric engine. At the same time, a square metal frame was placed over the conveyor belt. A Nikon D5200 digital camera was placed in the center of the metal frame facing downward. Then, a volume of olives and residues was placed in the small tank to simulate the conveyor belt mounted on the harvester. The olive fruits running over the conveyor belt were recorded in video format by the camera. The frames from all videos were extracted as images using a python script. The videos were recorded under natural conditions of daylight and included vibration, occlusion, and overlapping of both olive fruits and residues. A total of 1,000 images were annotated in the Visual Object Classes (VOC) format (Everingham et al., 2010), using the LabelImage tool (Tzutalin, 2015). The transfer learning technique to reuse an existing pre-trained model (Faster R-CNN) originally trained on the COCO data set was used. The "data augmentation" technique was also implemented during training. The model was trained during 20,000 steps, achieving a loss value of 0.21.

To deploy the trained model, a Coral USB Accelerator and Google Coral Dev Board were used. These devices allow to perform powerful AI-enabled offline computation in real-time processing (Natarov et al., 2020) with an attached RGB camera. The Coral Board with the RGB camera was placed in the same place where the Nikon D5200 was attached. Then, a known volume of olives with residues was put into the tiny tank. The olives and residues over the conveyor belt were detected with an accuracy of 95% and 70%, respectively.

The test in the prototype is just a preliminary study in laboratory conditions. To test the model in field conditions, it is intended to install a box with a similar metal frame over the conveyor belt of a straddle harvester (New Holland Braud 9090X Olive, CNH Global, Zedelgem, Belgium). This box will be equipped with the Coral Board, an RGB camera, and a cooling system to maintain the hardware under operative temperatures. Moreover, in the next steps, a mathematical model will be developed to relate the number of olive fruit to fruit mass and its suitability to be sent and displayed on a monitoring screen.

ACKNOWLEDGMENTS

This project has been carried out in collaboration with Agroplanning Agricultura Inteligente S.L. and within the Smart Biosystems Laboratory research group.

WORKS CITED

Ahmad, L., & Mahdi, S. S. (2018). Yield monitoring and mapping. In L. Ahmad and S. S. Mahdi (Eds.), *Satellite farming* (pp. 139–147). Springer.

Bazame, H. C., Molin, J. P., Althoff, D., & Martello, M. (2021). Detection, classification, and mapping of coffee fruits during harvest with computer vision. *Computers and Electronics in Agriculture, 183*, 106066.

Boatswain Jacques, A. A., Adamchuk, V. I., Park, J., Cloutier, G., Clark, J. J., & Miller, C. (2021). Towards a machine vision-based yield monitor for the counting and quality mapping of shallots. *Frontiers in Robotics and AI, 8*, 41. https://doi.org/10.3389/frobt.2021.627067

Chinchuluun, R., Lee, W. S., & Ehsani, R. (2009). Machine vision system for determining citrus count and size on a canopy shake and catch harvester. *Applied Engineering in Agriculture, 25*(4), 451–458.

Colmenero-Martinez, J. T., Blanco-Roldán, G. L., Bayano-Tejero, S., Castillo-Ruiz, F. J., Sola-Guirado, R. R., & Gil-Ribes, J. A. (2018). An automatic trunk-detection system for intensive olive harvesting with trunk shaker. *Biosystems Engineering, 172*, 92–101.

Chung, S. O., Choi, M. C., Lee, K. H., Kim, Y. J., Hong, S. J., & Li, M. (2016). Sensing technologies for grain crop yield monitoring systems: A review. *Journal of Biosystems Engineering, 41*(4), 408–417.

Everingham, M., Van Gool, L., Williams, C. K., Winn, J., & Zisserman, A. (2010). The pascal visual object classes (voc) challenge. *International Journal of Computer Vision, 88*(2), 303–338.

Kamilaris, A., & Prenafeta-Boldú, F. X. (2018). Deep learning in agriculture: A survey. *Computers and Electronics in Agriculture, 147*, 70–90.

Natarov, R., Dyka, Z., Bohovyk, R., Fedoriuk, M., Isaev, D., Sudakov, O., ... & Langendörfer, P. (2020, June). Artefacts in EEG signals epileptic seizure prediction using edge devices. In *2020 9th Mediterranean conference on embedded computing (MECO)* (pp. 1–3). IEEE.

Pérez-Ruiz, M., Rallo, P., Jiménez, M. R., Garrido-Izard, M., Suárez, M. P., Casanova, L., ... & Morales-Sillero, A. (2018). Evaluation of over-the-row harvester damage in a super-high-density olive orchard using on-board sensing techniques. *Sensors, 18*(4), 1242.

Tzutalin. (2015). LabelImg. Git code. https://github.com/tzutalin/labelImg

5.4 SMART AGRICULTURE BASED ON CYBER-PHYSICAL SYSTEMS

Diego Francisco Larios Marín and Francisco Javier Molina Cantero

ABSTRACT

The general objective of the project is to develop a cyber-physical system based on low-cost portable multispectral IoT nodes and Artificial Intelligence that allows the most relevant parameters that influence olive quality to be measured in a simple and economical way, with the benefit of assisting the farmer in making a decision about when to collect the olive from the olive tree and its processing for production in both table olives and olive oil.

INTRODUCTION

This project proposes the creation of a monitoring system based on artificial vision techniques, using multispectral images and AI algorithms applied to crops of products derived from the olive tree. The aim is to obtain precise information on those organoleptic variables of olives that have an impact on the quality parameters of the olive oil obtained from them. The monitoring is carried out at two different moments in the process: in the field, during the last months of the campaign, and in the oil mill, when the harvested olives are received.

The system provides information on the level of maturity, the fat yield, and the acidity of the olives. These data, traditionally obtained through laboratory analysis, are obtained through IoT nodes. The nodes incorporate a processing system that analyzes the image and calculates the corresponding values of the quality parameters in real time.

Data from these samplings are integrated into a software platform that processes and combines them to offer the farmer precise information on the state of the crop that helps them make logistical decisions (i.e., optimal date of collection) and management (irrigation/fertilizers), in order to achieve a higher production while maintaining the quality of the oil.

STATE OF THE ART

The use of hyperspectral images in the field of support for decision-making in agriculture has been considered in the literature by different authors. In most practical applications, it is the infrared bands that provide the most relevant information for classification and pattern detection tasks. In the specific case of products derived from the olive tree, there are various proposals for the evaluation of characteristics, such as nitrogen and potassium in olive grove plots (Gómez-Casero, M.T. et al., 2007) or acidity, humidity, and peroxide value in olive oil samples (Martinez-Gila et al., 2015). The use of spectrometry techniques (Vis/NIR) directly in the fruit has been proposed by various authors (León et al., 2003), (Cayuela et al., 2009) for the generation of predictive models of quality parameters: moisture, dry matter, oil content, free acidity, and maturity index. These predictive models use different wavelength sets for each parameter.

Once the hyperspectral information has been acquired and calibrated, a system is needed for the execution of complex algorithms and the evaluation of the characteristics of the crops to be determined. For these algorithms to work properly, it is important to have a large number of samples that contain both positive and negative aspects, that is, as many objects that contain what is intended to be measured, as well as objects that do not, in order to define what are the characteristics of the reflection of the mentioned object. In order to achieve this, it is important to have a large collection of previous images, in a catalog that facilitates application development.

Solving this in traditional computing structures is expensive and complex since it generally requires oversizing above the expected demand. For this reason, the use of technologies based on the use of cyber-physical systems is proposed, which, as stated in the European Digital Strategy (Cyber-Physical European Roadmap & Strategy, 2013-21015), is a disruptive technology that allows controlling and coordinating processes, on both local and global scale, using communication and information technologies.

CONTRIBUTION

The project has been developed by the TIC150 research group of the University of Seville in collaboration with the company SOLTEL, an Andalusian ICT consultancy. The stages in which the project has been developed are briefly detailed below.

a) *Development of a cataloging and labeling tool*

Initially, the need to develop a tool for cataloging olive samples was identified, so that an adequate knowledge base could be built for the analysis and training of pattern detectors. Subsequently, integration, validation, preprocessing, and semi-automatic labeling functions were added.

b) *Automatic image validation*

It is a fundamental function for the creation of the knowledge database, for the algorithm training activities, and also for the effectiveness of the real-time detectors. The factors that most influence a correct interpretation of the scene are the lighting and the distance between the camera and the olive tree. The perspective from which the scene is captured also influences, although to a lesser extent.

For the validation process, a reference framework implemented in Teflon is used in the capture process. The developed algorithm looks for that frame within the scene. From the spectral analysis of the detected Teflon, it is possible to deduce the characteristics of the ambient light, in order to correct its influence on the rest of the scene. Using a fixed lens, the size of the frame in the scene allows us to ensure that the distance between the camera and the scene is similar in all captures. This aspect is very important for two reasons. The first reason is to guarantee that the size of the olives is sufficient to be detected, and the second one is to be able to compare the performance in terms of the number of fruits based on their density and not on their number within the scene.

c) *Generation of a hyperspectral signature database*

Having a database with the spectral characteristics of the different analysis objectives is essential to obtain efficient classifiers.

For this purpose, the project has implemented an algorithm that processes cataloged images in batches. The algorithm performs the validation of the scene for each image, as well as the extraction of spatial scales and lighting correction parameters. The corrected mean spectrum is calculated for the area marked as olive and it is associated with the labeled degree of maturation. The spectra of all olives with the same ripening index are averaged to obtain a base signature. The most frequent univocal spectral signatures present in the same degree of maturation are calculated.

Using this technique, the knowledge model will be based not on one but on a set of signatures. This method facilitates the interpretation of the models and increases the effectiveness and speed of detection of the classifiers.

d) *Classifier training*

In this work, the paradigm known as "Supervised Learning" has been chosen, where each example is associated with a discrete label that identifies its membership in a group or class. The set of examples is used both to train the algorithms and to validate the results of their learning. The learning objective may be the classification of new examples or predictive inference.

Once the algorithm has been trained, it is capable of performing several functionalities: olives detection, olives counting, maturity index estimation, pest detection, and pest growth.

e) *Integration*

Data integration has required coordination with the other partners to define the information format and the exchange mechanism between the applications involved in the capture, labeling, and analysis of the images. A distributed architecture based on REST services has been chosen for information exchange.

The intelligent analysis module executes the extraction of information and the classification and estimation algorithms in real time, that is, as users upload the images to the web platform. The algorithms for preprocessing, information extraction, classification, counting, and estimation of the maturity index have been integrated into the management and operation system by means of a dynamic link DLL library that is called from the host that supports the application.

The computational performance is high since in the worst-case scenario the execution time is less than 3 seconds, if we exclude the upload times of the images (with sizes of 200 MB). Therefore, the viability of the developed algorithms is concluded.

ACKNOWLEDGMENTS

This project has been carried out in collaboration with SOLTEL IT SOLUTIONS and with the support of the following researchers: C. León, J. Barbancho, and J. I. Guerrero.

WORKS CITED

Cayuela, J. A., García Martos, J. M., & Caliani, N. (2009). NIR prediction of fruit moisture, free acidity and oil content in intact olives. *Grasas y Aceites, 60,* 194–202.

Cyber-physical European roadmap & strategy (2013–2015). https://cordis.europa.eu/project/id/611430/es

Gómez-Casero, M. T., López-Granados, F., Peña-Barragán, J. M., Jurado-Expósito, M., García Torres, L. & Fernández-Escobar, R. (2007). Assessing nitrogen and potassium deficiencies in olive orchards through discriminant analysis of hyperspectral data. *Journal of the American Society for Horticultural Science, 132*(5), 611–618.

León, L., Rallo, L., & Garrido-Varo, A. (2003). Near-infrared spectroscopy (NIRS) analysis of intact olive fruit: An useful tool in olive breeding programs. *Grasas y Aceites 54*(1), 41–47.

Martinez-Gila, D. M., Cano Marchal P., Gámez García, J., & Gómez Ortega, J. (2015). On-line system based on hyperspectral information to estimate acidity, moisture and peroxides in olive oil samples. *Computers and Electronics in Agriculture, 116,* 1–7.

5.5 DEVELOPMENT AND ADAPTATION OF DIGITAL INFORMATION MODELS FOR THE MANAGEMENT OF ARCHITECTURAL ASSETS: CASE STUDIES IN ARCHAEOLOGICAL AND ARCHITECTURAL HERITAGE

Francisco Pinto Puerto and Antonio García Martínez

ABSTRACT

The project develops and adapts digital models based on BIM (building information modeling) and GIS to the comprehensive and sustainable management of architectural heritage supported by CMMS (computerized maintenance management system) software. The experience of the research team and the involved companies and institutions provides the basis for creating processes and methodologies to adapt these models for use in two cases of architectural heritage: archaeological and social housing. The efforts are focused on defining a specific working methodology for each case through meetings with the companies concerned to establish their needs and requirements, create protocols, adapt resources, and redesign the workflows.

INTRODUCTION

Managing knowledge and the conservation and maintenance of important properties and architectures with heritage status is extremely time-consuming and resource-intensive for the institutions responsible for them. The daily tasks involved in inspection, simple conservation planning, restoration, and renovation require the acquisition and handling of vast quantities of information, all of which must be accessible in real time to ensure that the correct decisions are made. In this project, we explore the use of digital technological resources that can lead to much more efficient and sustainable management. To illustrate this, we have chosen two pilot cases that differ both in their architectures and, naturally, in the actions required to manage them. However, they have points in common regarding their complexity and the way in which the workflows are organized.

To carry out these experiences, we have created digital models that serve two purposes: steering the enormous body of data contributed from different disciplines (architecture, engineering, archaeology, history, geography, etc.) toward a convergence of knowledge and facilitating the rational application of practical results in the companies and institutions involved in the conservation and management of these architectural assets.

STATE OF THE ART

In recent years, numerous authors have pointed out the advantages of implementing BIM platforms to manage the use and maintenance of properties (Aziz, Nawawi & Ariff, 2016; Volk, Stengel & Schultmann, 2014).

The literature usually cites the main benefits of using these platforms, as follows: (a) efficient space management, (b) optimization of data storage and management, (c) common working and communication space for the involved agents, (d) possibility of controlling the social, environmental and economic impact of the life cycle, (e) efficient and preventive detection of the maintenance operations and necessary repairs, (f) ability to link the compiled data with other software programs, IoT, platforms, web resources, etc., (g) efficient maintenance of energy systems, and (h) optimized planning and operation of the common installations and services in buildings.

However, the transition from traditional management to more intelligent, efficient, and advanced BIM management requires the development of precise protocols and workflows. The benefits are particularly significant in the field of social housing, as shown by the results of studies conducted in Triolo (*Villeneuve d'Ascq*) in France and in the city of Wellington in New Zealand (Dunne, 2019).

The application of BIM to heritage architecture, known as HBIM, is giving rise to a growing body of scientific literature that examines very different aspects of its connection with digital information models (Mayowa et al., 2021; Pocobelli et al., 2018) In any case, HBIM can be considered to constitute a particular case of the BIM methodological baggage that accords special importance to the heritage status of the architecture in question.

The cultural significance of the properties where HBIM is applied highlights the problem of their geometric documentation. This must be carried out in terms of architectural survey; in other words, as a critical process consisting in gathering data about form and conducting an architectural analysis. This process can lead to different levels of knowledge about the concerned property.

The guardianship (comprehensive management) of architectural heritage encompasses identification, research, protection, and conservation. HBIM offers an efficient methodology for the sustainable management of architectural heritage, adaptable to different scales and types, as well as to the specific requirements of the agents responsible for its management, from documentation to obtaining legal protection to the registration and control of preventive conservation activities.

CONTRIBUTION

We set out with the aim of adapting BIM and GIS digital information systems to maintenance management supported by CMMS, in view of the broad and complex range of architectural heritage that needs to be maintained and preserved, so that it can continue to play a role in society. This particular project focuses on two specific cases: social housing for rent, which requires constant inspection and servicing by the Andalusian Housing and Renovation Agency (AVRA), and a large and complex set of highly vulnerable archaeological ruins that receive numerous cultural visits and are the topic of countless research projects—the Archaeological Site of Itálica (CAI). We have aligned the digital models to meet the needs specified by both institutions: for example, creating a workflow that facilitates the management of information for research purposes and improving the way information is obtained to assess the maintenance costs. This implies creating an open system to allow information to be constantly updated and processed so that users can create reports and tables or edit existing plans.

Our ultimate aim is to define the essential elements and the most appropriate workflow and create an inclusive system to facilitate use by agents who are not familiar with these tools.

We have, therefore, compared the current working dynamics of the two involved institutions, defined the elements that will be managed, and gathered data from the accumulated experience. We have also had to consider the human team and infrastructures available since in the short term they are the ones who are going to use the system.

We are currently undertaking four related actions:

Action 1. Codification of the assets and the conservation tasks based on the systems that the two institutions currently use: in the case of AVRA, the Building Evaluation Report (IEE), and in the case of the CAI the data tables, the graphical objects in CAD and the QGIS platform used to manage routine maintenance work. We are in the process of codifying the building systems and pathologies recorded in the AVRA databases for the public facilities it manages (approximately 70,000 housing units), and in the case of CAI for an archaeological site of approximately 45 hectares, plus various plots of land in the town of Santiponce and environs. In both cases, the aim is to standardize periodic preventive maintenance.

Action 2. Definition of an IT system, in the form of a toolbox, to contain the most common combinations and proposals of architectural and archaeological intervention for the subsequent comprehensive renovation or conservation project. The renovation will take into account physical, economic, and social parameters, as well as energy strategies in the case of AVRA, and in the case of the CAI, the conservation will include historical and artistic values, as well as its cultural appreciation.

Action 3. Integration into a collaborative BIM and GIS platform of all the involved agents, as well as the technologies necessary for the traceable and optimal digitalization of the information and the future management of the data, so that the pilot experience can be replicated in projects with similar general features managed by AVRA or received by the CAI. In the case of AVRA, a digital model is being generated for one of the properties and will incorporate selected information from the IEE. In the case of the CIA, one of the Roman houses and its immediate urban vicinity are being modeled, and we are also working on a selection of specific preventive conservation actions across the entire site.

Action 4. Creation of systems for transferring information between users from very different disciplines who will be responsible for managing the information and/or making the necessary operational decisions for the conservation, maintenance, and intervention in the affected architectures.

ACKNOWLEDGMENTS

This project has been carried out in collaboration with CAI and AVRA and with the support of the following researchers: Francisco Pastor Gil, Roque Angulo Fornos, José María Guerrero Vega y Manuel Castellano Román, and Heritage Knowledge Strategies research group.

WORKS CITED

Aziz, N. D., Nawawi, A. H., & Ariff, N. R. M. (2016). Building information modelling (BIM) in facilities management: Opportunities to be considered by facility managers. *Procedia—Social and Behavioral Sciences, 234*, 353–362. https://doi.org/https://doi.org/10.1016/j.sbspro.2016.10.252

Dunne, T. (2019). Applying BIM retrospectively as a data collection tool for maintaining social housing. *BIMinNZ case study: Wellington City Council Bracken Road Flats.* https://www.biminnz.co.nz/casestudies/2019/4/9/wellington-city-council-bracken-road-flats

Mayowa, I. A., Joseph, H. L., Edwin, H. C., & Amos, D. A. R. K. O. (2021). Heritage building maintenance management (HBMM): A bibliometric-qualitative analysis of literature. *Journal of Building Engineering*, 102416. https://doi.org/10.1016/j.jobe.2021.102416

Pocobelli, D. P., Boehm, J., Bryan, P., Still, J., & Grau-Bové, J. (2018). BIM for heritage science: A review. *Heritage Science*, 6(1). https://doi.org/10.1186/s40494-018-0191-4

Volk, R., Stengel, J., & Schultmann, F. (2014). Building information modeling (BIM) for existing buildings—Literature review and future needs. *Automation in Construction, 38*, 109–127. https://doi.org/10.1016/j.autcon.2013.10.023

Conclusion

As it has been pointed out, the importance of this book relies on the particular nature of the project "Innovative Ecosystem with Artificial Intelligence for Andalusia 2025", which has joined research groups from the University of Seville, on the one hand, and private sector, on the other. Together, these two parts have been working on the development and application of various solutions to the issues the different sectors currently face. Taking into account the fact that all these possible solutions are based on Artificial Intelligence techniques, the book is intended to be a type of manual or starting point for answering in an innovative way to different challenges Andalusia region faces in the health sector, sustainability, digital economy, mobility, agroindustry, and tourism (Figure C.1).

This means that the book, offering results obtained in 22 subprojects through the AI approach, is primarily oriented to R&D departments of both small- and medium-sized companies and large multinationals, as well as public institutions and health-care centers. These departments can find in this book essential information on how to establish a mutually beneficial relationship with universities and research centers in terms of knowledge transfer. The handbook might also help them understand the possibilities of using Artificial Intelligence in different sectors.

The book is also aimed at innovation centers, university communities, and research groups, as it is an example of best practices in knowledge transfer, university-business relations, and society and shows Artificial Intelligence technologies as a transversal opportunity for development and innovation in the different sectors shown in this handbook (health and social welfare, energy efficiency and sustainable construction, digital economy, mobility logistics and advanced industry linked to transportation, endogenous land-based resources, agroindustry, and tourism). It is also considered to be an opportunity for research groups to have a real impact on their environment, by enabling universities and research centers to be drivers of innovation and improvement of society and industry within their region.

Finally, it is aimed at university students, as it shows AI technologies as a transversal tool for the development of different disciplines, highlighting the importance for any student to have at least a basic knowledge of these technologies, whatever their area of study is.

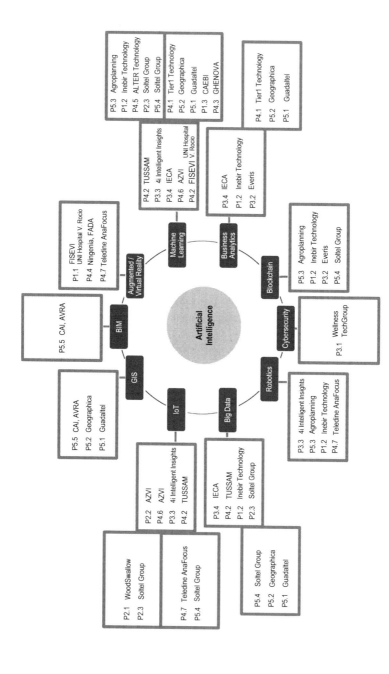

FIGURE C.1   Visual representation of AI technologies and companies or institutions involved in each sub-project.

Also, the book opens the opportunity to keep researching the field, especially taking into consideration the second edition of this project coordinated by the University of Seville, in which new companies and research groups will join in order to study the possibilities of applying Artificial Intelligence in solving different problems in today's Andalusian society and further.

José Guadix Martín
Milica Lilic
Marina Rosales Martínez

Index